Ecosocialism or barbarism

Socialist Resistance would be glad to have readers' opinions of this book, its design and translations, and any suggestions you may have for future publications or wider distribution.

Socialist Resistance books are available at special quantity discounts to educational and non-profit organizations, and to bookstores.

To contact us, please write to:
Socialist Resistance, PO Box 1109, London N4 2UU, Britain
or email us at: contact@socialistresistance.net
or visit: www.socialistresistance.net

Designed by Ed Fredenburgh
Published by Socialist Resistance, October 2006
Printed in Britain by Lightning Source
ISBN 0-902869-97-3
EAN 9780902869973

Ecosocialism or barbarism

Edited by Jane Kelly and Sheila Malone

Socialist Resistance, London

Publisher's Acknowledgements

This book appears only through the kind help of the authors. The permission of Monthly Review, Green Left Weekly, and Capitalism Nature Socialism is also gratefully recognised.
Endnotes are supplied by the authors; the footnotes are ours.

The articles in this volume were first published in the following periodicals:
Michael Löwy, 'What Is Ecosocialism?', Capitalism Nature Socialism, Vol 16 No 2, June 2005
Daniel Tanuro, 'Ecology: "Climate of Fear"', IVP, 358, April 2004
Sheila Malone, 'Kyoto's Answers to Climate Change' Socialist Outlook 6, 2005
Phil Ward, '"A large scale geophysical experiment?" Global warming, capitalism and our future', Socialist Outlook, 8, 2005
François Moreau, 'Environment: Our Common Future?', International Marxist Review, Summer 1990
Alice Cutler, 'The nuclear option: A solution to global warming?', Socialist Outlook, 6, 2005
Phil Ward, 'Nuclear Juggernaut Moving into Top Gear', Socialist Outlook 9, 2006
John Bellamy Foster, 'Organising ecological Revolution', Monthly Review, Vol 57, No 5, October 2005
Jane Kelly & Phil Ward, 'No Solution Under Capitalism', Socialist Outlook, 2, 2004
'Ecology and Socialism', International Viewpoint, Summer, 2003. Documents of the 15th World Congress of the Fourth international
Dick Nichols, 'Cuba's Green Revolution', Green Left Weekly, June 18, 2000. Quotations in that article are taken from Cuba Verde (Green Cuba), José Martí Publishing House, Havana, 1999.

Contents

Introduction

Socialist Resistance hopes to stimulate a socialist ecology that can unite and enrich both "reds" and "greens". The convergence of these movements could form a new vision for society – ecosocialism. The papers in this book were collected together for Socialist Resistance's forum on 'Ecosocialism or Barbarism' in London on December 2, 2006. It aims to bring together those who share a radical criticism of modern capitalism and the desire for a total alternative based on ecological and socialist practice. That alternative has to address at least four dimensions.

1 Solving ecological problems

There are big differences between scientific, reformist and revolutionary responses to ecological problems. This is clearest with the Kyoto protocols where the ruling class is trusted to resolve the crisis of climate change which it has itself created. However, struggles across the Middle East have also shown that capitalist war increasingly attacks the environment. Reformists have also failed agricultural regions. They are being ruined, driven by massive increases in the use of fertilizers and poisons. Nor has the green movement responded effectively to the challenge that genetics now gives businesses the tools to produce new organisms and biological weapons that have unknowable dangers.

2 What politics are needed for what ecology?

Over the last decade many ecologists have devoted their energies to making business greener. This approach encourages capitalists to think green when they design, manufacture and use natural resources and energy. It reflects a strong ideological project to convince activists that ecologically-sustainable capitalism is possible. In turn, it has produced an "ecology industry" which recycles green campaigners into defenders of industrial capitalism. This reformist ecology needs to be subjected to a profound criticism. It often reflects the conservative "new age" approach that many ecologists have in which ecology is counter-posed to humanity. Rather ecology must be fought for as a human right: to water, to energy, to stable climates and to enjoy life.

3 Popular ecology in the 'South'

The limited success of the green parties in Europe is very different to the massive popularity of ecological struggles in Latin America. In Cuba, Venezuela, Brazil, Bolivia and elsewhere, mass movements have attempted to turn around the ecological devastation of capitalism. Today, Cuba is in many ways an example to the world of the gains produced by popular ecology. Ecology isn't a 'Western luxury', as Chico Mendes and other Brazilian campaigners discovered when confronted with the murderous violence of the capitalist class. Across Latin America, capitalism's pillage of environmental resources has provoked a massive response from workers and farmers.

4 Growth, nature and abundance

Ecosocialism also aims to re-examine the dominant idea of Europe's "economistic" left that still equates progress with eternally increasing production, without considering the impact on future generations. Some socialists repeat the capitalists' lie that happiness comes from the unlimited abundance of goods, rather than from transformed human relations. The ecologist movement also faces virulent debates, especially about overpopulation and the influence of economic, political and ideological campaigns of the most polluting companies. Under socialism we can get more from less: a democratically planned economy will make more useful things with fewer resources and less waste.

The growing impact of climate change around the world makes the integration of ecology and socialism a pressing task. Our hope is that this book, like the forum, brings "reds" and "greens" closer together.

October, 2006.

Satellite photo of flooding of the Upper Elbe, 2002

Floods reach upper Elbe
NOAA 17 2002-08-22
www.fvalk.com

Part one: solving ecological problems

What Is Ecosocialism?

Michael Löwy

Translated by Eric Canepa

The reigning capitalist system is bringing the planet's inhabitants a long list of irreparable calamities. Witness: exponential growth of air pollution in big cities and across rural landscapes; fouled drinking water; global warming, with the incipient melting of the polar icecaps and the increase of "natural" extreme weather-related catastrophes; the deterioration of the ozone layer; the increasing destruction of tropical rainforests; the rapid decrease of biodiversity through the extinction of thousands of species; the exhausting of the soil; desertification; the unmanageable accumulation of waste, especially nuclear; the multiplication of nuclear accidents along with the threat of a new— and perhaps more destructive—Chernobyl; food contamination, genetic engineering, "mad cow," and hormone-injected beef.

All the warning signs are red: it is clear that the insatiable quest for profits, the productivist and mercantile logic of capitalist/industrial civilization is leading us into an ecological disaster of incalculable proportions. This is not to give in to "catastrophism" but to verify that the dynamic of infinite "growth" brought about by capitalist expansion is threatening the natural foundations of human life on the planet.

How should we react to this danger? Socialism and ecology— or at least some of its currents—share objective goals that imply a questioning of this economic automatism, of the reign of quantification, of production as a goal in itself, of the dictatorship of money, of the

1

reduction of the social universe to the calculations of profitability and the needs of capital accumulation. Both socialism and ecology appeal to qualitative values—for the socialists, use-value, the satisfaction of needs, social equality; for the ecologists, protecting nature and ecological balance. Both conceive of the economy as "embedded" in the environment—a social environment or a natural environment.

That said, basic differences have until now separated the "reds" from the "greens," the Marxists from the ecologists.Ecologists accuse Marx and Engels of productivism. Is this justified? Yes and no. No, to the extent that no one has denounced the capitalist logic of production for production's sake—as well as the accumulation of capital, wealth, and commodities as goals in themselves—as vehemently as Marx did. The very idea of socialism—contrary to its miserable bureaucratic deformations—is that of production of use-values, of goods necessary to the satisfaction of human needs. For Marx, the supreme goal of technical progress is not the infinite accumulation of goods ("having") but the reduction of the working day and the accumulation of free time ("being"). Yes, to the extent that one often sees in Marx and Engels (and all the more in later Marxism) a tendency to make the "development of the productive forces" the principal vector of progress, along with an insufficiently critical attitude toward industrial civilization, notably in its destructive relationship to the environment.

In reality, one can find material in the writings of Marx and Engels to support both interpretations. The ecological issue is, in my opinion, the great challenge for a renewal of Marxist thought at the threshold of the 21st century. It requires that Marxists undertake a deep critical revision of their traditional conception of "productive forces," and that they break radically with the ideology of linear progress and with the technological and economic paradigm of modern industrial civilization.

Walter Benjamin was one of the first Marxists in the 20th century to articulate this question. In 1928, in his book *One-Way Street*, he denounced as an "imperialist doctrine" the idea of the domination of nature and proposed a new conception of technology as a "mastery of the relations between nature and humanity." Some years later in *On the Concept of History*, he proposed enriching historical materialism with the ideas of Fourier, that utopian visionary who dreamt of "labor,

2

which, far from exploiting nature, would be capable of awakening the creations that slept in its womb."

Today Marxism is still far from having made up for its backwardness in this regard. Nevertheless, certain lines of thinking are beginning to tackle the problem. A fertile trail has been opened up by the American ecologist and "Marxist-Polanyist," James O'Connor. He proposes that we add to Marx's first contradiction of capitalism—that between the forces and relations of production—a second contradiction—that between productive forces and conditions of production—which takes into account workers, urban space, and nature. Through its expansionist dynamic, O'Connor points out, capital endangers or destroys its own conditions, beginning with the natural environment—a possibility that Marx did not adequately consider.

Another interesting approach is one suggested in a recent piece by the Italian "ecomarxist," Tiziano Bagarollo: "The formula according to which there is a transformation of potentially productive forces into effectively destructive ones, above all in regard to the environment, seems more appropriate and meaningful than the well-known scheme of the contradiction between (dynamic) forces of production and relations of production (that are fetters on the former). Moreover, this formula provides a critical, non-apologetic, foundation for economic, technological, and scientific development and therefore the elaboration of a "differentiated" (Ernst Bloch) concept of progress."

Whether Marxist or not, the traditional labor movement in Europe—unions, social-democratic and communist parties—remains profoundly shaped by the ideology of "progress" and productivism, even leading it, in certain cases, without asking too many questions, to the defense of nuclear energy or the automobile industry. However, ecological sensitivity has begun to emerge, notably in the trade unions and left parties of the Nordic countries, Spain, and Germany.

The great contribution of ecology has been, and still is, to make us conscious of the dangers threatening the planet as a result of the present mode of production and consumption. The exponential growth of attacks on the environment and the increasing threat of the breakdown of the ecological balance constitute a catastrophic scenario that calls into question the survival of the human species. We are facing a crisis of civilization that demands radical change. The problem is that the proposals put forward by the leading circles of European political ecology are, at best, highly inadequate and at worst, wholly inappropriate solutions to the ecological crisis. Their main weakness is that they do not acknowledge the necessary connection between productivism and capitalism. Instead, reforms like eco-taxes capable of controlling "excesses" or ideas like "green economics" lead to the illusion of a "clean capitalism." Or, further, taking as a pretext the imitation of Western productivism by bureaucratic command economies, they conceive of capitalism and "socialism" as variants of the same model—an argument that has lost a lot of its attraction after the collapse of so-called "actually existing socialism."

Ecologists are mistaken if they imagine they can do without the Marxian critique of capitalism. An ecology that does not recognize the relation between "productivism" and the logic of profit is destined to fail—or, worse, to become absorbed by the system. Examples abound. The lack of a coherent anti-capitalist posture led most of the European Green parties—notably, in France, Germany, Italy, and Belgium— to become mere "eco-reformist" partners in the social-liberal management of capitalism by center-left governments. Regarding workers as irremediably devoted to productivism, certain ecologists have avoided the labor movement and have adopted the slogan "neither left nor right." Ex-Marxists converted to ecology hastily say "goodbye to the working class" (Andre Gorz), while others (Alain Lipietz) insist on the need to abandon "the red"—that is, Marxism or socialism—to join "the green," the new paradigm thought to be the answer to all economic and social problems.

Finally, in so-called "fundamentalist," or deep-ecology circles, we see, under the pretext of opposing anthropocentrism, a rejection of humanism, which leads to relativist positions that place all living species on the same plane. Should one really maintain that Koch's bacillus or the Anopheles mosquito have the same right to life as a child suffering from tuberculosis or malaria?

What then is ecosocialism? It is a current of ecological thought and action that appropriates the fundamental gains of Marxism while shaking off its productivist dross. For ecosocialists, the market's profit logic, and the logic of bureaucratic authoritarianism within the late departed "actually existing socialism," are incompatible with the need

4

to safeguard the natural environment. While criticizing the ideology of the dominant sectors of the labor movement, ecosocialists know that the workers and their organizations are an indispensable force for any radical transformation of the system as well as the establishment of a new socialist and ecological society.

Ecosocialism developed mostly during the last 30 years, thanks to the work of major thinkers like Raymond Williams, Rudolf Bahro (in his earlier writings) and Andre Gorz (also in his early work), as well as the very useful contributions of James O'Connor, Barry Commoner, John Bellamy Foster, Joel Kovel, Joan Martínez-Alier, Francisco Fernández Buey, Jorge Riechman (the latter three from Spain), Jean-Paul Déléage, Jean-Marie Harribey (France), Elmar Altvater, Frieder Otto Wolf (Germany), and many others, who publish in journals like *Capitalism Nature Socialism* and *Ecología Politica*. This current is far from politically homogeneous. Still, most of its representatives share certain common themes. Breaking with the productivist ideology of progress—in its capitalist and/or bureaucratic form—and opposed to the infinite expansion of a mode of production and consumption that destroys nature, it represents an original attempt to connect the fundamental ideas of Marxian socialism to the gains of critical ecology.

James O'Connor defines as ecosocialist the theories and movements that seek to subordinate exchange-value to use-value, by organizing production as a function of social needs and the requirements of environmental protection. Their aim, an ecological socialism, would be an ecologically rational society founded on democratic control, social equality, and the predominance of use-value. I would add that this conception assumes collective ownership of the means of production, democratic planning that makes it possible for society to define the goals of investment and production, and a new technological structure of the productive forces. Ecosocialist reasoning rests on two essential arguments:

1. The present mode of production and consumption of advanced capitalist countries, which is based on the logic of boundless accumulation (of capital, profits, and commodities), waste of resources, ostentatious consumption, and the accelerated destruction of the environment, cannot in any way be extended to the whole planet

without a major ecological crisis. According to recent calculations, if one extended to the whole world the average energy consumption of the United States, the known reserves of petroleum would be exhausted in nineteen days. Thus, this system necessarily operates on the maintenance and aggravation of the glaring inequality between North and South.

2. Whatever the cause, the continuation of capitalist "progress" and the expansion of a civilization based on a market economy—even under this brutally inequitable form in which the world's majority consume less—directly threatens, in the middle term (any exact forecast would be risky), the very survival of the human species. The protection of the natural environment is thus a humanist imperative. Rationality limited by the capitalist market, with its short-sighted calculation of profit and loss, stands in intrinsic contradiction to ecological rationality, which takes into account the length of natural cycles.

It is not a matter of contrasting "bad" ecocidal capitalists to "good" green capitalists; it is the system itself, based on ruthless competition, the demands of profitability, and the race for rapid profit, which is the destroyer of nature's balance. Would-be green capitalism is nothing but a publicity stunt, a label for the purpose of selling a commodity, or—in the best of cases—a local initiative equivalent to a drop of water on the arid soil of the capitalist desert. Against commodity fetishism and the reified autonomisation of the economy brought about by neoliberalism, the challenge of the future for ecosocialists is the realization of a "moral economy." This moral economy must exist in the sense in which EP Thompson used this term, that is, an economic policy founded on non-monetary and extra-economic criteria. In other words, it must reintegrate the economic into the ecological, the social, and the political.

Partial reforms are completely inadequate; what is needed is the replacement of the micro-rationality of profit by a social and ecological macro-rationality, which demands a veritable change of civilization. That is impossible without a profound technological reorientation aimed at the replacement of present energy sources by other non-polluting and renewable ones, such as wind or solar energy.

The first question, therefore, concerns control over the means of production, especially decisions on investment and technological change, which must be taken away from the banks and capitalist enterprises in order to serve society's common good. Admittedly, radical change concerns not only production but consumption as well. However, the problem of bourgeois/industrial civilization is not—as ecologists often assert—the population's "excessive consumption." Nor is the solution a general "limit" on consumption. It is, rather, the prevalent type of consumption, based as it is on ostentation, waste, mercantile alienation, and an accumulationist obsession, that must be called into question. An economy in transition to socialism, "re-embedded" (as Karl Polanyi would say) in the social and natural environment, would be founded on the democratic choice of priorities and investments by the population itself, and not by "the laws of the market" or an omniscient politburo. Local, national, and, sooner or later, international, democratic planning, would define what products are to be subsidized or even distributed without charge; what energy options are to be pursued, even if they are not, in the beginning, the most profitable; how to reorganize the transportation system according to social and ecological criteria; and what measures to take to repair, as quickly as possible, the enormous environmental damage bequeathed to us by capitalism. And so on...

This transition would lead not only to a new mode of production and an egalitarian and democratic society, but also to an alternative mode of life, a new ecosocialist civilization, beyond the reign of money, beyond consumption habits artificially produced by advertising, and beyond the unlimited production of commodities, such as private automobiles, that are harmful to the environment. Utopia? In its etymological sense ("nowhere"), certainly. However, if one does not believe, with Hegel, that "everything that is real is rational, and everything that is rational is real," how does one reflect on substantial rationality without appealing to utopias?

Utopia is indispensable to social change, provided it is based on the contradictions found in reality and on real social movements. This is true of ecosocialism, which proposes a strategic alliance between "reds" and "greens"—not in the narrow sense used by politicians applied to social-democratic and green parties, but in the broader sense between the labor movement and the ecological movement—

and the movements of solidarity with the oppressed and exploited of the South. This alliance implies that ecology gives up any tendency to anti-humanist naturalism and abandons its claim to have replaced the critique of political economy.

From the other side, Marxism needs to overcome its productivism. One way of seeing this would be to discard the mechanistic scheme of the opposition between the forces of production and the relations of production, which impede them. This should be replaced—or at least, be completed—by the idea that productive forces in the capitalist system become destructive ones. Take, for example, the armament industry, or the various branches of production that are destructive of human health and of the natural environment.

The revolutionary utopia of green socialism, or of solar communism, does not imply that one ought not to act right now. Not having illusions about "ecologizing" capitalism does not mean that one cannot join the battle for immediate reforms. For example, certain kinds of eco-taxes could be useful, providing they are based on an egalitarian social logic (make the polluters pay, not the public) and that one disposes of the economic-calculation myth of "market price" for ecological damages, which are incommensurate with any monetary point of view. We desperately need to win time, to struggle immediately for the banning of the HCFCs that are destroying the ozone layer, for a moratorium on genetically modified organisms, for severe limitations on the emissions of greenhouse gases, and to privilege public transportation over the polluting and anti-social private automobile.

The trap awaiting us here is the formal acknowledgement of our demands, which empties them of content. An exemplary case is that of the Kyoto Protocol on Climate Change, which provides for a minimal reduction of 5 percent of gases responsible for global warming in relation to 1990—certainly too little to achieve any results. As is known, the US, the main power responsible for the emission of these gases, has stubbornly refused to sign the protocol. As for Europe, Japan and Canada, they have signed the protocol while adding clauses such as the famous "market of rights of emission," which enormously restrict the treaty's already limited reach. Rather than the long-term interests of humanity, it is the short-term view of the oil multinationals and the

8

automobile industry that has predominated.

The struggle for ecosocial reforms can be the vehicle for dynamic change, a "transition" between minimal demands and the maximal program, provided one rejects the pressure and arguments of the ruling interests for "competitiveness" and "modernization" in the name of the "rules of the market."

Certain immediate demands have already, or could rapidly, become the locus of a convergence between social and ecological movements, trade unions and defenders of the environment, "reds" and "greens"; the promotion of inexpensive or free public transportation—trains, metros, buses, trams—as an alternative to the choking and pollution of cities and countrysides by private automobiles and the trucking system; the rejection of the system of debt and extreme neoliberal "structural adjustments" imposed by the International Monetary Fund and the World Bank on the countries of the South, with dramatic social and ecological consequences: massive unemployment, destruction of social protections, and destruction of natural resources through export; the defense of public health against the pollution of the air, water and food, due to the greed of large capitalist enterprises; and the reduction of work time to cope with unemployment and create a society that privileges free time over the accumulation of goods.

All of the emancipatory social movements must be brought together to birth a new civilization that is more humane and respectful of nature. As Jorge Riechmann says so aptly: "This project cannot reject any of the colors of the rainbow—neither the red of the anti-capitalist and egalitarian labor movement, nor the violet of the struggles for women's liberation, nor the white of non-violent movements for peace, nor the anti-authoritarian black of the libertarians and anarchists, and even less the green of the struggle for a just and free humanity on a habitable planet."

Radical political ecology has become a social and political force present on the terrain of most European countries, and also, to a certain extent, in the US. However, nothing would be more wrong than to regard ecological questions as only of concern to the countries of the North—a luxury of rich societies. Increasingly, social movements with an ecological dimension are developing in the countries of peripheral capitalism—the South. These movements are reacting to a growing

9

aggravation of the ecological problems of Asia, Africa and Latin America that result from a deliberate policy of "pollution export" by the imperialist countries.

The economic "legitimation"—from the point of view of capitalist market economy—was bluntly articulated in an internal World Bank memo by the institution's chief economist, Lawrence Summers (formerly president of Harvard University) in *The Economist* in early 1992. Summers said: "Just between you and me, shouldn't the World Bank be encouraging more migration of dirty industries to the LDCs [less developed countries]? I can think of three reasons:

1. The measurement of the costs of health-impairing pollution depends on the forgone earnings from increased morbidity and mortality. From this point of view a given amount of health-impairing pollution should be done in the country with the lowest cost, which will be the country with the lowest wages. I think the economic logic behind dumping a load of toxic waste in the lowest wage country is impeccable, and we should face up to that.

2. The costs of pollution are likely to be non-linear as the initial increments of pollution probably have very low costs. I've always thought that under-populated countries in Africa are vastly under-polluted; their air quality is probably vastly inefficiently low compared to Los Angeles or Mexico City...

3. The demand for a clean environment for aesthetic and health reasons is likely to have very high income-elasticity. The concern over an agent that causes a one-in-a-million change in the odds of prostate cancer is obviously going to be much higher in a country where people survive to get prostate cancer than in a country where under-five mortality is 200 per thousand..." In this statement we see a cynical formulation that clearly reveals the logic of global capital—in contrast to all the mollifying speeches on "development" produced by the international financial institutions. In the countries of the South, we thus see the birth of movements which Joan Martínez-Alier calls "the ecology of the poor" or even "ecological neo-narodnism." These include popular mobilizations in defense of peasant agriculture, communal access to natural resources threatened with destruction by the aggressive expansion of the market (or the state), as well as struggles against the degradation of the local environment caused by unequal

exchange, dependent industrialization, genetic modifications and the development of capitalism (agribusiness) in the countrysides.

Often these movements do not define themselves as ecological, but their struggle nevertheless has a crucial ecological dimension. It goes without saying that these movements are not against the improvements brought by technological progress; on the contrary, the demand for electricity, running water, sewage, and more medical dispensaries are prominent in their platforms. What they reject is the pollution and destruction of their natural surroundings in the name of "market laws" and the imperatives of capitalist "expansion."

A recent article by the Peruvian peasant leader Hugo Blanco is a striking articulation of the meaning of this "ecology of the poor": "At first sight, the defenders of the environment or the conservationists appear as nice people, slightly crazy, whose principal aim in life is to prevent the disappearance of blue whales and panda bears. The common people have more important things occupying them, for example, how to get their daily bread. However, there are in Peru a great many people who are defenders of the environment. To be sure, if one told them "you are ecologists," they would probably reply "ecologist my eye." And yet, in their struggle against the pollution caused by the Southern Peru Copper Corporation, are not the inhabitants of the town of Ilo and the surrounding villages defenders of the environment? And is not the population of the Amazon completely ecologist, ready to die to defend their forests against pillage? The same goes for the poor population of Lima, when they protest against water pollution."

Among the innumerable manifestations of the "ecology of the poor," one movement is particularly exemplary, by its breadth, which is at once social and ecological, local and global, "red" and "green": the struggle of Chico Mendes and the Forest People Alliance in defense of the Brazilian Amazon against the destructive activity of the large landowners and of multinational agribusiness. Let us briefly recall the main aspects of this confrontation.

In the early 1980s, a militant trade-unionist linked to the Unified Workers Confederation (CUT) and partisan of the new socialist movement represented by the Brazilian Workers Party, Chico Mendes organized occupations of land by peasants who earned their livelihoods from rubber tapping (seringueiros) against the latifundistas who

11

bulldozed the forest in order to establish pasture lands. Later Mendes succeeded in bringing together peasants, agricultural workers, seringueiros, trade unionists and indigenous tribes—with the support of the church's base communities—to form the Forest People Alliance, which blocked many attempts at deforestation. The international outcry resulting from these actions earned him the United Nations Global 500 Prize in 1987. But shortly afterwards, in December of 1988, the latifundistas made him pay dearly for his struggle by having him killed by paid assassins.

Through its linking of socialism and ecology, peasant and indigenous struggles, survival of local populations and taking responsibility for a global concern (the protection of the last great tropical rain forest), this movement can become a paradigm of future popular mobilizations in the South. Today, at the turn of the 20th century, radical political ecology has become one of the most important ingredients of the vast movement against capitalist neoliberal globalization, which is developing in the North as well as the South.

The massive presence of ecologists was one of the striking aspects of the big demonstration in Seattle against the World Trade Organization in 1999. And at the World Social Forum in Porto Alegre in 2001, one of the most powerful symbolic acts of the event was the operation led by activists of the Landless Movement and José Bové's French Farmer's Confederation: the digging up of a field of Monsanto's genetically modified corn.

The battle against the uncontrolled spread of genetically modified food is mobilizing in Brazil, France and other countries. This struggle brings together not only the ecological movement but also the farmers' movement, part of the left, and members of the general public who are disturbed by the unforeseeable consequences of genetic modification on public health and the natural environment. The struggle against the commodification of the world and the defense of the environment, resistance to the dictatorship of multinationals, and the battle for ecology are intimately linked in the reflection and praxis of the world movement against capitalist/liberal globalization.

Kyoto: climate of fear

Daniel Tanuro

Heat waves, droughts and floods have focused attention on climate change caused by the accumulation of greenhouse gases. Governments try to reassure us that, whether the Kyoto Protocol is ratified or not, adequate measures will continue to be taken and the problem will be brought under control. The reality is, alas, much more worrying.

Even George W Bush does not dare to argue the point: "There is a natural greenhouse effect that contributes to warming. Greenhouse gases trap heat, and thus warm the earth because they prevent a significant proportion of infrared radiation from escaping into space. Concentrations of greenhouse gases, especially CO_2, have increased substantially since the beginning of the industrial revolution. And the National Academy of Sciences indicates that the increase is due in large part to human activity."[1] For two centuries, deforestation, industry and transport have led to the accumulation in the atmosphere of gases that admit sunlight to the earth but prevent the earth's infrared rays from reaching space. The result is that, as in a greenhouse, the atmosphere heats up. Average temperatures increased by 0.6°C in the 20th century – a growth unprecedented for nearly 10,000 years – bringing about an increase in ocean levels of 10 to 25 cm. The process is accelerating and if nothing changes, the Inter-government Panel on Climate Change (IPCC) predicts a global warming of between 1.4 and 5.8°C by 2100, leading to a rise in water levels of 9 to 88 cm.[2] Global warming caused by human activity has certainly begun, and it is irreversible. It affects not only the atmosphere, but also gigantic masses of oceanic waters; since the inertia of these latter is considerable, the process will make its effects felt for at least a thousand years.

The social, economic and environmental consequences are incalculable. Detailing them is not the aim of this article. Nonetheless, let us recall the strong words of John Houghton, former chief executive of the British Meteorological Office and co-president of the "Scientific Evaluation" Working Group of the IPCC:

"Global warming is now a weapon of mass destruction. It kills more people than terrorism, yet Blair and Bush do nothing".[3] Humanity must try to bring about stabilization at a new point of equilibrium. It is in this context that the Kyoto Protocol – concluded in 1997 in the context of the United Nations Framework Convention on Climate Change – signified the resolve of the developed countries to reduce their emissions of greenhouse gases by 5.2% on average over the period 2008-2012, the year 1990 serving as a benchmark. The European Union set itself the objective of an 8% reduction.[4] Six years after its negotiation in Japan, and in spite of its signature by 119 countries, the Protocol is stymied. In order to be applied, it should be ratified by 55 states accounting for more than 55% of emissions. However, the main world producer of greenhouse gases withdrew in 2001 – the USA refused to sign any agreement that did not impose commitments on the big developing countries like China and India. Washington was also opposed to what it saw as a weakening of the competitiveness of a highly polluting US energy sector largely built on oil and coal.[5] In this context, ratification by Moscow became indispensable.[6] But Vladimir Putin used the situation to raise the stakes in relation to Europe and Japan, who were partisans of an agreement. Andrei Illarionov, the President's main economic adviser, declared recently that the Protocol went against Russia's national interests.[7] In this arm wrestling match, the key point is the price of a tonne of carbon. Depending on the latter, the gains that Russia and the Ukraine could make in selling their emission credits would vary between 20 and 170 billion dollars in five years. Indeed the price would be higher if the US signed the agreement, for it is that country that faces the most difficulty in conforming to Kyoto.[8] While a pure and simple abandonment is not to be ruled out, it is probable that Kyoto will survive this game of poker. But it will be a still more neutered Kyoto, since the Protocol is being undermined from the inside by forces which use Russia's reticence and US rejection as pretexts to reduce their demands, indeed to conceal their inability to meet them.[9]

What will be the future without the Protocol? Some experts are reassuring: "The treaty has already changed the world in small but significant ways that will be hard to reverse", says the *New York Times*:

14

"From Europe to Japan and the United States, just the prospect of the treaty has resulted in legislation and new government and industry policies curbing emissions."[10] This optimism is misplaced, for four reasons.

The limitations of the Protocol

First, whatever its neoliberal inspiration, Kyoto has the advantage of posing a double constraint: figures for objectives of reduction, and a timetable. These aspects are in the firing line of certain industrial lobbies and their political spokespersons: "If global warming turns out to be a problem, which I doubt, it won't be solved by making ourselves poorer through energy rationing. It will be solved through building resiliency and capability into society and through long-term technological innovation and transformation", according to Myron Ebell, climatic specialist with the Competitive Enterprise Institute. Former assistant secretary of State and Kyoto specialist in the Clinton administration, David B Sandalow, says, "The standard of success isn't whether the first treaty out of the box sails through. The standard is whether this puts the world on a path to solving a long-term problem. Other multilateral regimes dealing with huge complex problems, like the World Trade Organization, have taken 45 or 50 years to get established."[11]

Secondly, whether the Protocol is amended or abandoned, the measures taken will be much less than the initial objectives – and the latter were already completely insufficient. According to the IPCC, by 2050 emissions should be lowered not by 5. 2% but by 60% in order that average warming does not exceed 2°C in relation to the pre-industrial era. As for the timetable, if it is true that climate change is a very long-term process, it does not follow that humanity can wait 50 years in order for industry to adapt its capacity without reducing its profits. On the contrary, the more the measures are delayed and limited, the longer any return to a point of equilibrium will take, the higher this point of equilibrium will be and the more serious will be the consequences.[12] The climate constitutes what mathematicians call a "complex chaotic system" – limited changes can make it cross qualitative thresholds leading to rapid upheavals.[13]

15

The projections on temperature and water levels give an indication of urgency. The IPCC estimates that the former could climb from 1.4 to 5.8°C by 2100. On both sides of these planetary averages, there are – by definition – extremes. Above Greenland, for example, warming is one to three times the world average. Even on the lowest prediction of the experts, it is then possible that this region would gain 2 to 3°C in 50-80 years. An increase of this magnitude would be enough to melt the Greenland icecap in a few centuries, which would lead to a rise in sea levels of 6 metres.[14]

Thirdly, the USA wants the big developing countries to carry a part of the climatic burden – a demand which appears unjust to the countries of the South. The figures speak for themselves: "To stabilize levels of greenhouse gases at a level twice those at the time of the industrial revolution, global emissions would have to be reduced from the current one tonne of carbon per person per year to an average of 0.4 tonnes", says Larry Lohman. "The US emits 13 times this amount per head, or 5.2 tonnes, and Japan and Western European nations five to twelve times this amount per head, or two to five tonnes. More than 50 Southern countries including India, by contrast, emit less than half the maximum level, or 0.2 tonnes per person".[15] Certainly, all countries should adopt a responsible attitude in relation to the climate. But it is in the developed world that it is necessary to start, unilaterally, while massive technological aid should be provided to the countries of the South. The US demand amounts to saying that the dominated nations should pay for the changes in climate of which their populations are the first victims; these changes have been caused above all by 200 years of capitalist development in the North, at the price of the pillage and non-development of the rest of the planet.

Fourthly, the small step taken at Kyoto implies perverse effects, often little known. These relate to two issues, that of "carbon sinks", on the one hand, and the commodity logic of "flexible mechanisms" on the other. With or without the Protocol, these categories will play a growing role as alternatives to the "energy rationing" mentioned by Mr Ebell. These questions should then be subjected to a broad public debate.

Carbon sinks

Culprit number one for the growth of the greenhouse effect is carbon dioxide gas, which plays a major role in the carbon cycle. Schematically, the process is as follows:

1. CO_2 is absorbed by green plants which, thanks to chlorophyll and sunlight, transform it into cellulose;
2. this transformation is known as photosynthesis;
3. the closing of the cycle takes place through respiration and the decomposition of dead organisms, which liberate the carbon contained in organic matter (in the form of CO_2 or methane).

However, there is carbon dioxide gas and carbon dioxide gas. The burning of oil, coal or natural gas brings new quantities of carbon into the cycle and as plants (and soils and oceans) cannot absorb it completely, a part of this carbon accumulates in the atmosphere (mainly in the form of CO_2), increasing the greenhouse effect.[16]

From the viewpoint of the struggle against climate change, it is then vital to distinguish between two very different processes. On the one hand, reduction at source of emissions originating from the use of fossil fuels and on the other the reduction of the concentration of atmospheric CO_2 due to absorption by green plants (we speak in this case of the "capture" of carbon in "sinks"). The first aspect is strategically decisive. The IPCC tells us that it is practically certain that emissions of CO_2 due to the burning of fossil fuels will remain the determinant factor in the evolution of the concentration of CO_2 in the atmosphere during the 21st century.[17] Carbon sinks will only allow an attenuation of climatic change.[18]

However, this hierarchy between the "determinant factor" and the "attenuating factors" is not found in the Kyoto Protocol. On the contrary, the Protocol amalgamates reduction of emissions and increase of absorptions, and invites the states to balance the two processes.[19] In other words, planting enough trees, or cultivating without tillage, would allow the burning of oil to continue. It is a short-term logic, whose perverse effect is obvious – the problem is put off to future generations, while continuing to get worse. It is also a wrong headed logic, because it is difficult to measure exactly the net quantity of CO_2 absorbed by the ecosystems, or to predict the evolution of this absorption in relation to global warming and the growing concentration of CO_2.[20]

The commodity logic of "flexible mechanisms"

Three different mechanisms are embodied in the Protocol – "joint implementation", "clean development" and "emissions trading". All three have as their goal the lightening of the economic cost of the commitments made at Kyoto.

"Joint implementation" allows the developed countries that have signed the protocol to attain their objectives of reduction of emissions through common investment. In Europe, for example, companies from the West who make investments in the East to increase energy efficiency can make proportional adaptations to their own emission levels. It is enough for them to "prove" that emissions would have been more significant if the investment had not been made. In this respect, the substitution of natural gas for coal as the source of electricity production opens vast possibilities to foreign companies and governments. Research consultancies are specializing in the identification of these opportunities. Thus the Norwegian group Point Carbon puts Romania at the top of its hit parade – no country is better placed to welcome joint implementation projects, it chortles.[21] The perverse effect is that these investments in the East (which would have taken place anyway, in the context of the buying up of the "New Europe" by western capital) allow the big industrial groups in the West to drag their feet on more complicated and costly modern installations and technological adaptations which are indispensable to the struggle against the greenhouse effect.

The "clean development mechanism" allows a developed state to make an investment in a country in the South that reduces emissions (or increases absorptions), and to correspondingly adapt its own levels of emission. In this framework, the EU is multiplying its efforts to sell clean technologies to the countries of the South. Better still, some polluting enterprises buy land in the Third World, plant rapid growth trees and thus acquire carbon credit corresponding to the CO_2 that they create by burning fossil fuels in the North. Inside the

18

EU, Holland is the champion of this neo-colonial practice, followed by Finland, Austria and Sweden.[22] But US big business is not far behind and, with or without Kyoto, companies are conscious that measures against climatic change are inescapable. They want to position themselves for future climate negotiations, to take their share of the market and improve their brand image among consumers. In this respect, what better than to participate in projects of reforestation in the Third World? It suffices to hide the negative effects, which are numerous – these "industrial plantations of trees" (they are not "forests"!) accelerate the rural exodus and the decline of food-producing cultures, accentuating dependence on exports and re-colonization, and damaging ecosystems and biodiversity (see section at end of this article on the Plantar project in Brazil). And do not forget that the "clean development mechanism" does not reduce pollution in the North – on the contrary, it allows it to continue, with its attendant consequences for health and the environment.

The spectre of air privatization

Tradeable emissions are the cornerstone of the "flexible mechanisms". Each signatory country is given emission quotas. The developed states divide up these quotas among the companies established on their territory. Those which remain below their objective can sell "rights to pollute" to others. Whether Kyoto is ratified or not, all the big polluting companies are involved in these exchanges – in the US credits are sold on the Chicago Climate Exchange.[23] According to some economists, at the price of $14 per tonnes of carbon, the "emission rights" created in the framework of Kyoto would lead to a carbon trade corresponding to the abstract creation of 2,345 billion dollars, or the biggest creation of monetary capital through an international treaty in history.[24] A system of exchange has already been set up inside the EU, whereby from 2005 onwards "clean" enterprises, can sell their pollution credits to "dirty" companies (5,000 big companies are already involved at the first stage). Here also, as in the case of "joint implementation" the East constitutes a veritable reservoir of carbon. Big consumers of energy before the fall of the Berlin Wall, the economies of the former "Soviet bloc" subsequently collapsed. Owing to the reference date for their objectives within the framework of Kyoto, these countries dispose of "carbon credits" that other signatories can acquire, thus avoiding reductions in their own emissions.

From the viewpoint of the market economy it is not absurd to manage the reduction of the production of greenhouse gases in this way. The system of emission rights has functioned in the USA to reduce the rates of sulphur dioxide (SO_2) in the air, and thus acid rain. The ecological efficacy of the system depends on political will, which is expressed in the establishment of quotas and the rhythm of their decline. However, recourse to this kind of mechanism requires a broad debate within society, for the commodification of the emission rights could take humanity towards an outcome which is at first blush unthinkable: the privatization of the air. Some argue that "selling the wind" is and will remain impossible. But doesn't buying the "right to pollute" affirm ownership over the thing polluted? In the precise case of carbon dioxide gas, the question is not absurd given that, once discharged in the atmosphere,[25] CO_2 cannot easily be separated from the other components of air – nitrogen, oxygen and so on. To be the owner of millions of tonnes of "waste CO_2," is certainly equivalent to being the "owner" of polluted atmospheric masses. Certainly, air cannot be materially enclosed. However, its legal enclosure is perfectly possible. The countries of the North and their companies that are awarded emission quotas will be inclined to consider these as semi-permanent ownership rights. The dividing up of these rights, the result of 200 years of imperialist development, will tend to be considered as the "natural" proportion of shares of the atmosphere attributed to different countries and groups of countries. In the developed countries, legal arrangements could rapidly be imposed on citizens to make them pay for the "services" of the atmosphere or, at least, for the cost of its de-pollution. The maintenance of these services will be entrusted to the private sector and, in the name of the environment, the true cost will be imposed on consumers (as in the water sector), while the companies will benefit from competitive prices. As for the countries of the South, they would be victims of a kind of "climatic neo-colonialism". If they wish to increase their emission quotas, they will be accused of irresponsibility;

to develop, they will be forced to buy clean technologies from Northern companies. Moreover, they will be confronted with the fact that these companies, thanks to the Clean Development Mechanism, will have snapped up the "carbon sinks" and other cost-effective means of compensating for their emissions.[26]

Sorcerers' apprentices: progress towards ratification of the Kyoto Protocol, as of 1 January 2003

If the negotiations around the Kyoto Protocol are so arduous, it is because everyone knows henceforth that climate change is a reality that will necessitate very considerable adaptations. Such measures will interfere in the US-EU-Japan relationship of forces, to the point that climate change becomes a major geo-strategic issue. The neoliberal offensive around "flexible mechanisms" and "sinks" takes place in this context. Its function is of course to weaken and indeed head off an indispensable and urgent reduction in the source of emissions of greenhouse gases, with the goal of protecting the rate of profit of the big companies. But the offensive also seeks, more broadly, to make the struggle against climate change a profitable business (see below), an instrument of domination of the South and a new frontier in the capitalist drive to appropriate natural resources.

We have to note that this offensive is gaining ground and at the Conferences of Parties (COP) to the UN Framework Convention discussions on "flexible mechanisms" increasingly take precedence over the reduction in emissions originating from the use of fossil carbon. At the COP-9 in Milan in December 2003 amendments presented by Norway, seeking to ban monocultures and GMOs, were rejected. While climate change becomes increasingly palpable and menacing, the crazy logic of capitalist accumulation draws humanity more and more towards a major catastrophe.

The market in non-polluting forms of energy is a theatre of ferocious economic competition. This clarifies the role of the main protagonists in the climatic negotiations, the US and the EU.

Deprived of significant sources of oil and cheap natural gas, confronted with US domination of the Middle East and traumatized by Chernobyl, the EU is developing an energy policy based on the diversification of supplies, growing energy efficiency and the development of renewable energies. Currently, 6% of the energy used in the EU comes from renewable sources,[27] and the objective is to reach 12% in 2010.[28]

Indeed, such a strategy necessitates big public investments, in the form of aid for research, economic incentives and public sector orders, the goal being to support companies in the initial stage.

Given globalization and the opening up of new markets, these efforts are only sustainable if the relative prices of energy originating from renewable sources become competitive in relation to those of energy produced from fossil fuels, if use of the latter is restrained, and if a world market in "clean" technologies is opened (the three conditions being linked). Kyoto involves a response at different levels. With the Kyoto protocol in place, the world market in clean technologies should prosper, according to a document of the Commission.[29] The stakes are significant. The world market in the eco-industry is estimated at 550 billion euros. The experts count on its enlargement in the five coming years, above all in the emergent countries, with growth rates of 5 to 8%.[30]

The EU is well placed to play an important role. Its companies are in world leading positions in the sector of renewable energies, notably in the construction of windmills (75% of the planet's current capacity). It is easy to understand why the EU, far from being shaken by the clash with the US during the negotiations at the Hague in November 2000, held out until the agreements of Marrakech and Kyoto, then launched a "Coalition for Renewable Energy" – "the OPEC of Renewables", according to the Commissioner for the Environment – which now involves eighty countries.

The US approach is less monolithic than the European press has tended to imply. The powerful environmentalist lobby has some influence on the Republican Party, to the extent that the adoption of norms for emissions of greenhouse gases figured in Bush's electoral programme in 2000![31] More fundamentally, the world of business is divided. "The prospect of other countries moving ahead with limits on greenhouse gases while the US sticks its head in the sand worries many American companies. With the evidence that human activities are causing global warming getting ever more convincing, emissions curbs in many countries are inevitable, execs believe" "Economies will have to adjust to that," says Tom Jacob, manager for international and industry affairs at DuPont. "It would be a mistake if the US economy is insulated from those pressures. When the reality comes, the US will have a bigger game of catch-up – and our competitors will be ahead of us in developing and using climate-friendly technologies."[32] On either side of the Atlantic, then, ecological concerns do not weigh too heavily on the "climatic" strategies which are being followed.

The Plantar project[33] in Minas Gerais (Brazil) is a good example of the ravages of the kind of "reforestation" carried out in the framework of Clean Development Mechanisms (CDM).

Developed under the auspices of the World Bank, Plantar is an industrial plantation of eucalyptus in monoculture (23,100 hectares) destined for the production of charcoal for the steel industry. It is also the first "carbon sinks" project to seek to register with the CDM Executive Board, the international body which is responsible in this area.

According to the documentation provided by Plantar, the project will allow the production of 3.8 million tonnes of steel products in 21 years as well as the creation of 3,000 jobs. But Plantar has met violent opposition, for social reasons (low wages, insecure jobs) and ecological reasons (massive use of the herbicide glyphosate, exhaustion and contamination of water resources, killing of fish, significant reduction of biodiversity).

Eucalyptus grows very quickly, and Plantar has committed itself to maintaining its "carbon sinks" for 42 years. If the status of CDM is accorded to it, the enterprise will serve as "compensation" for the emission into the atmosphere of millions of tonnes of CO_2 which will certainly not disappear as quickly as the trees.

Through the Prototype Carbon Fund of the World Bank, Plantar benefits from the support of three European governments (Sweden, Holland and Finland), Gaz de France and the Belgian company Electrabel.

Kyoto's Answers to Climate Change

Sheila Malone

Climate change is a time bomb ticking away in our world, already ravaging lives and communities, and threatening catastrophes against which most of us will have little protection.

The UN's Intergovernmental Panel on Climate Change predicts that the earth's temperature will rise by anything from 1.4 to 5.8 percent by the end of this century, due overwhelmingly to our dependency on fossil fuels as our main energy source. Greenhouse gases are being poured into the atmosphere at the staggering and unsustainable rate of seven billion tons annually. Emissions have risen by 20 percent in the last 50 years, by 10 percent just in the last decade, and are set to rise still more dramatically in the future.

The result has been to upset the delicate climatic balance of our planet, leading to increasing extremes in weather conditions. The European heat wave of 2003 and the freak floods in Carlisle last year are just examples that affect Britain. As temperatures continue to rise and ice caps melt, coastal and low-lying areas will experience severe flooding, others drought and desertification and the spread of tropical diseases like malaria. Increased turbulence from the released energy will produce more and more violent storms, hurricanes and tornadoes. And as the seas soak up emissions they will gradually acidify, harming plankton, which are at the bottom of the food chain. The worst-case scenario is that climate change could spiral out of control, making the earth uninhabitable for humans.

Aware of the seriousness of the issue, most governments and politicians, as well as big business, are increasingly keen to be seen tackling the problem and displaying their 'green credentials'. Our own Prime Minister Tony Blair has recently pledged to make climate change a priority issue, and a flurry of paperwork is already emerging from Downing Street. This includes the UK Sustainable Development Strategy, as well as a new government service 'Environment Direct', which is all about consumer choice in pollution.

Typically New Labour, most of these initiatives are long on words and short on funding. In fact, at the same time as mouthing this sort of 'greenwash' the government is actually cutting back on support for environmentally friendly projects – for instance, it has recently stopped subsidies for solar panel installation. It is also forging ahead with huge road and airport expansion schemes. Carbon dioxide emissions continue to increase under this government, by 2.2 percent in 2003 and again by 1.2 percent in 2004 (DTI figures, *Guardian* 01.04.05). It will therefore fail to meet even its mild reduction target under the Kyoto Protocol.

The Kyoto Protocol

Governments and business are making much of this recently agreed and first legally binding global treaty on climate change. Kyoto commits thirty-four of the most industrialised nations to an overall cut in CO_2 emissions by five percent of 1990 levels by the year 2012. This is hardly anything to get excited about, since most climatologists agree that in fact a reduction of 60 percent by at least 2050 is needed. Also the biggest polluter of all, the USA, responsible for 25 percent of global greenhouse gas emissions has refused to sign up to the treaty at all, saying this would damage its economy.

Actually, the US's present stance is unlikely to continue. This is because of the central method for implementing Kyoto, that is the creation of a market to trade in carbon emissions – an idea first thought up by the USA itself, and potentially much too lucrative to miss out on.

The Carbon Emissions Trading Scheme (CETS) is a mechanism thoroughly in line with imperialism's neo-liberal policies. It promotes the idea that, just as for everything else, the unfettered 'free' market is the solution to climate change. Pollution is turned into a commodity from which to make profit.

Under the CETS, countries who have agreed to the cuts targets are given 'emissions credits' to pollute within that limit. They can either use up the whole allowance, 'bank' some credits for the future or sell them to another polluter on the open market. If a polluter wants to exceed their allowance they can buy spare credits from another or earn more through investing in so-called pollution reduction schemes

25

such as the Clean Development Mechanism.

Under Kyoto therefore, corporate capitalism will not only be able to continue to pollute with minimal restriction, but also be able to make handsome profits. Significantly, the UK has already spent £215 million on its trial trading scheme, in anticipation of the money to be made. Kyoto's new carbon market enshrines the domination of the already big and powerful multinationals. Not surprisingly, they have some thoroughly imperialist schemes for 'offsetting' pollution in the North onto the poor countries of the South.

With World Bank assistance, hundreds of thousands of hectares of land in third world countries are to be given over to 'carbon sinks' – vast tree plantations to soak up CO_2 gases. In addition to their questionable science, these displace communities, destroy local agriculture and create dependency, just as externally imposed monocultures have done in the past. In Brazil, the World Bank is funding a eucalyptus plantation and as Heidi Bachram argues,

While plantations have their own ecologically destructive qualities such as biodiversity loss, water table disruption and pollution from herbicides and pesticides, their social impact is equally devastating to a local community. Lands previously used by local people are enclosed and in some cases they have been forcibly evicted.... The workers on such plantations have little or no health and safety protection and are exposed to hazardous chemicals and dust particles.... Similar disregard exists for the natural environment. Thus the fisher folk in the regions around the plantations in Brazil are poverty-stricken and devastated due to the pollution caused by over-use of pesticides and herbicides, which contaminates rivers and water sources and kills fish. In some cases, the water in streams and rivers is entirely dried up because the non-indigenous eucalyptus is a thirsty tree.[34]

Nevertheless, companies like Future Forest are already selling branded offset products for so-called Carbon Neutral living. You can now buy Carbon Neutral motoring and air travel and live in a Carbon Neutral home, courtesy of this carbon dumping on the poor.

Other schemes involve 'geological sequestration' or the pumping of CO_2 emissions back into old oil wells. The US and Australian governments are each investing around $90 million a year in research into this unproven and risky enterprise. But enthusiasm is great, since the pressure used enables companies to squeeze out at least another 10 per cent of the fast diminishing oil. Given these profitable offsetting schemes, and until fossil fuels finally run out, it actually pays energy companies to carry on polluting, rather than invest in renewable sources. Consumers in the rich North are also led to believe they need not change their polluting lifestyles.

Limitless exploitation of capitalism

Capitalism operates on the basis that the earth's resources are there for limitless exploitation, and that market forces will always find a (benign) solution to a crisis. In fact their solutions to the growing energy crisis are increasingly violent. In the South it has meant more and more wars for control of decreasing resources, such as the savage war on Iraq to gain control of its oil. And in the North, under the guise of the spurious 'war on terror' governments are also putting in place repressive legislation, which can conveniently deal with dissent and protest when major ecological disasters begin to occur.

There has always been resistance in colonised countries to subjugation and pillage by imperialism and its local elites, whether to military conquest or economic devastation through monocultures, deforestation, huge dam building and so on. Opposition by indigenous peoples has often been based on an understanding of the need to respect the limits of nature's resources in order to survive.

Our 24-hour fossil fuel driven economies want us to deny this. But it is at our peril. The 90 percent cuts in CO_2 emissions needed by mid century to halt the escalating effects of global warming cannot be done without a massive and immediate switch to investment in renewable energy sources (solar, wave, wind), combined with widespread and comprehensive conservation measures, such as building insulation, more and better recycling and so on.

Campaigning for such measures changes our attitude to and behaviour towards our environment, both individually and collectively. This is important if we are to work out and plan a necessary alternative economy and way of life. Individually, it is estimated that

27

we are each responsible for 10 tons of CO_2 emissions annually. We need information, advice and financial help from our governments on how we can help in reducing this amount now. The alternative is just to carry on dumping the consequences of our profligacy onto the poor in developing countries and onto future generations.

When fossil fuels finally run out, the energy multinationals will be forced to make a switch to renewables. However, in the meantime, the most recent report of the UN Millennium Ecosystems Assessment, the work of 1,300 scientists in 95 countries, estimates that we have already used up two thirds of the earth's resources, irreversibly damaging ecosystems. Capitalism's profit-driven overproduction and over-consumption is proving unable to halt this catastrophe.

Production for use, not profit

Social and ecological movements have increasingly joined together to campaign for an alternative. Many struggles on health and food focus on the high human cost of pollution. Others campaign for better and affordable public transport, instead of the overuse of the private car. There is also a debate about what kind of renewable energy sources would best replace fossil fuels, for example wind or solar power. Some in the environmental movement have even raised the issue of using nuclear power. (See Alice Cutler, 'A nuclear solution to global warming?' in this volume, for arguments against this.)

Whatever the switch, the important issue is the rejection of the idea that economic growth must mean the maximum exploitation of human and natural resources. This is the 'productivism' of capitalism that has led to commodification and alienation in our relation to our environment, but which is presented as the only model of development, whether in the rich North, the former Soviet bloc or in the developing countries of the South.

An alternative, socialist model would mean wresting control of our economies from profit-hungry corporate capitalism. Social ownership and control could then begin to plan for human need, instead of private gain.

28

'A large scale geophysical experiment?': Global warming, capitalism and our future

Phil Ward

Capitalism has known about global warming for at least fifty years but seems unable to face up to the consequences. Is there something inherent in capitalism that prevents its moving to a low Greenhouse Gas economy? Although planning under workers' and community control is what we hope to see in the future, there are also demands we should be making on governments now.

While the idea that some of the gases in the atmosphere helped to warm the earth was first propounded by Joseph Fourier as early as 1824, by 1896, the science behind global warming was broadly understood and it was predicted that a doubling of concentrations of the greenhouse gas (GHG) CO_2 would warm the earth by 4-5°C. It was also shown that such a doubling was possible through the burning of fossil fuels.

Measurements showing a rise in average global temperatures and carbon dioxide concentrations since 1860 were first made in 1939 and confirmed in the 1950s. By 1965, the US President's Science Advisory Committee said that the projected 25 percent increase in CO_2 by 2000, 'may be sufficient to produce measurable and perhaps marked changes in climate'. By then, most of the climate sceptics' arguments about the oceans being able to absorb the excess gas, or temperatures not having risen since 1940 had been dealt with.

In 1989, *Scientific American* announced that climate scientists had reached a consensus on climate change, and called for a 50 percent cut in global fossil fuel consumption and a stop to deforestation. This call followed the finding that the 1980s had experienced the six hottest years on record (back to 1000 CE). By 2000, the 1990s were the warmest decade on record. Now, the six warmest years have been 1998, 2005, 2002, 2003, 2001 and 2004.[35]

Global warming has been having profound effects on world

ecosystems for at least 20 years. Already in 1989, it was reported that arctic permafrost and sea ice were retreating, glaciers receding and the average temperatures in the great lakes had risen. In 2005, further evidence on all these phenomena was published: the Siberian permafrost was shown to be melting, threatening the release of millions of tonnes of trapped methane, 99 percent of Alaska's two thousand low altitude glaciers were shown to be retreating, the great lakes were shown to be thawing two days earlier per decade since 1846 and freezing later.

Now new measurements and models have increased the reliability of past temperature measurements, putting warming data on a firmer footing. Methane (a more powerful GHG than CO_2) was found to contribute to a third of global warming, rather than one sixth. Even if all GHG emissions stopped immediately, 'a potentially dangerous level of global warming cannot be ruled out', due to the time taken for sea temperatures to respond to current GHG levels.[36] A study of Antarctic ice cores showed that CO_2 levels are 30 percent higher and methane levels three times higher than at any time in the last 650,000 years. The increasing acidity of the oceans (due to more dissolved carbon dioxide) was found to be a threat to ocean life and one of the North Atlantic sea currents is 30 percent weaker now than fifty years ago.

Finally, it has been shown that in the last fifty years, the destructive power of hurricanes has increased by 70 percent: this is now attributed to global warming by many scientists. Typically, the sceptics on this issue refer mainly to the ten percent of hurricanes that occur in the Atlantic, where the evidence is less clear-cut, ignoring the 90 percent in the Indian and Pacific oceans.[37]

Other likely outcomes of climate change have not yet been recorded. These include a major disruption of agriculture, resulting in more widespread water shortages and famine (of course exacerbated by late capitalist globalisation). This could happen not only in the so-called 'third world' but also in the 'West', where agribusiness is very rigid and inadaptable, while climate change is more extensive nearer the poles. One other effect causing concern is the creation of huge, unstable meltwater lakes in glacial areas. These lakes are likely to burst and inundate downstream regions, as was recently reported in Bhutan. Other consequences could be the spreading of tropical diseases to temperate zones and unpredictable effects on human survival capacity in what is now recognised to be a major extinction event that is taking place.

The statement, made as early as 1957, by a climatologist that, 'human beings are now carrying out a large scale geophysical experiment of a kind that could not have happened in the past', could not be more apt.

So capitalism has had at least fifty years to respond to the threat of global warming. Its wholly inadequate response should lead us to question whether there is something inherent in globalised market capitalism that prevents its moving to a low GHG economy. We must look beyond issues like Bush's resistance to Kyoto[38] and Montreal and engage with the underlying systemic problems.[39] One approach is to start with what kind of policies would be necessary to get massive cuts in GHG emissions.

What policies do we need?
The role of renewable energy resources

While there is clearly a place for all kinds of renewable technologies, and some of these could be put into use relatively quickly, George Monbiot[40] has shown that they cannot replace fossil fuels, even at today's levels of energy usage, let alone for the future. Official US government figures project a 60-80 percent increase in energy demand between now and 2025, with China's demand almost tripling. Wind power may supply up to 20 percent of current UK electricity demand (replacing nuclear power), but that is only 6 percent of total *energy* usage, which includes gas for heating and petrol for transport as well. And Britain has the best wind resources in Europe.

Monbiot has also highlighted the ripping up of virgin rain forest in SE Asia to create oil palm monocultures to feed Asia's cooking pots and Europe's cars. He quotes Friends of the Earth saying that this demand for biomass causes 85 percent of Malaysia's forest loss, acknowledged but still encouraged by the UK government. The World Commission on Dams has questioned the role of hydroelectric power as a renewable resource: some dams contribute as much to global warming as equivalent fossil fuel power stations, and they all damage habitats and fisheries and displace people.

Running the world's motor vehicle fleet on biofuels is not feasible: one calculation suggests that biomass for US vehicles would take up 97 percent of the total land area of that country (mountains, deserts and all), while another shows that to run all the world's cars would take 60 percent of the land currently devoted to agriculture. Hydrogen power is not an option either, as large amounts of energy are required to generate the gas from water. Hydrogen also destroys the ozone layer in the atmosphere[41].

Sequestration

Renewable energy resources have their own environmental downside. The same applies to carbon dioxide storage (sequestration) from fossil fuel power stations, now under consideration. World CO_2 emissions are about 26bn tonnes per year. Even putting in place the technology to store 10 percent of this would be a major undertaking fraught with environmental hazards. It is also energy-intensive, requiring either the liquefying and distillation of air (to facilitate burning in pure oxygen), or the heating and cooling of noxious chemicals (amines) to separate CO_2 from nitrogen after combustion. In any case, new, untried technologies will take too long to put into place.

Nuclear power

These considerations have led to increasing calls for nuclear power to be used to replace fossil fuels. Again, there is a problem with the time frame, the extent to which nuclear can replace other energy resources and with the potential costs. There is the political problem that governments have to underwrite the liabilities of nuclear power companies – no insurance company is willing to provide full cover. No responsible government should ever burden its citizens or future generations with the hazards that nuclear power brings.

However it looks as if New Labour will go for the nuclear option. The British nuclear generator BNFL now owns the main plant manufacturer, Westinghouse, which may explain Blair's enthusiasm. Wind power, which has been shown to be more feasible that nuclear, is based mainly on German or Danish technology. [42]

Reduction of global demand

In the face of all the drawbacks for alternative energy sources, the central policy in reduction of GHG emissions has to be a reduction in world demand for energy. Some of this reduction could be met by conservation measures like home insulation, and we should be demanding government subsidies to upgrade and insulate Britain's housing stock, as well as making sustainable energy such as wind and solar power cheap and available to all. But the bulk will have to come about by lowering the level of economic activity, at least in the imperialist countries. Here lies the rub, at least for the capitalist system, which is incapable of downsizing except by means of destructive slump or war.

Developing countries

There will also need to be a major change in the trajectory of developing countries. One major issue is the migration of people from rural areas to mega-cities. China illustrates this most starkly. Here, expected migration is over 400m people in the coming years. *Newsnight* reporter Paul Mason has suggested that, even if the whole of Western manufacturing industry were transferred to China, there would still not be enough work in the cities for these people.

There is the added problem of the power demands of mega-cities – China is currently commissioning one coal-fired power station a fortnight. The biggest airports, stadiums, tower blocks and road systems are today being constructed in China. It has sixteen of the twenty most polluted cities and 400,000 people die prematurely of respiratory diseases as a result. Pollution has doubled there in the last ten years and could quadruple again by 2020. In November and December 2005, there were three major incidents of water pollution threatening to poison people in large cities. The Chinese model of 'development', reminiscent of that of England in the 1800s, so graphically described by Engels, is clearly unsustainable. In many respects, it is a model replicated all over the developing world. The people of the developing countries have the right to escape grinding poverty and backwardness, but this must be achieved in a sustainable way, by social planning nationally and globally. Globalising capitalism shows itself increasingly incapable of this.

Most of the developing countries have greater scope for renewable energy sources than the west, at least in rural areas. Solar power would usually be more efficient: biomass, currently the number one energy source, can be much more efficiently used, by altering agricultural practices and access to stoves. The political priority is a shifting of development from prestige mega-projects in the cities, to improving the quality of life in the countryside and small towns. The centre of this would be land reform, the building of collective political structures and the empowerment of women.

Our demands

For the urban areas in developing countries, the priorities are essentially the same as for the imperialist powers, including a diminution of the economy. The impending catastrophe is global, and requires global, national and local solutions. International treaties need to go far beyond Kyoto. The fight must be taken up on all fronts. The kinds of national and international measures required are as follows:

- A huge energy conservation programme, freely available home insulation and new designs for buildings so they need minimal heating and cooling;
- A transport policy that reduces our need to use cars;
- Less use and therefore production of motor cars and lorries;
- Development of alternative production;
- Localisation of production and consumption wherever possible;
- The planning of towns and cities so that public transport is efficient and people do not have to travel far to access shops, libraries and entertainment;
- Shortening of the working week and increasing holidays, making it feasible to use public transport for commuting and trains as an alternative to air travel;
- Rationing of air travel for all;
- Promotion of more communal living situations leading to a

reduced production of consumer durables, the setting up of neighbourhood childcare facilities and cafés, laundries;

- A sharp reduction in meat consumption, which is having disastrous effects in Brazil, Argentina and other countries, as well as causing major GHG emissions and adding to pressure on water supplies;
- The development of international plans on the use of water, to deal with interstate water conflicts such as between Iraq, Syria and Turkey;
- The development of measures to deal with the increasing areas of both drought and floods;
- Curtailing of activities not essential to human well-being, such as the advertising, sales, arms and many other industries;
- Ending competition between firms – leading to many firms producing the same commodities. Working people should decide what is produced based on human need, not profit, as well as the environmental impact the production and use of goods has;
- In less-developed countries, policies for developing rural areas are necessary. Land reform is essential and a focus on production for local use rather than the world market. Also needed are renewable energy sources (particularly solar) where there is no grid; measures to ensure water conservation; safe use of sewage in areas without a sewage system, etc.

Many of these necessary measures would not be compatible with capitalism, which demands continuous growth in order to secure its profits. Meeting these demands implies that we need a planned economy not left to the anarchy of market forces, which has got us into our current mess.

These types of policy imply radical change in social relations which explains the difficulty the capitalist class and their political representatives have in coming up with measures to combat climate change. Furthermore, at present it is perfectly rational for an individual firm to ignore climate change. For individual firms, the logic of capitalism is the production of the maximum exchange value, regardless of the effects of their commodities on the environment. However, the accumulated 'micro-rationality' of thousands of firms adds up to the macro-irrationality of climate change that will threaten the capitalist system.

Mass action

Ultimately of course, governments may be *forced* to act against climate change. The best situation would be if this resulted from a mass movement demanding effective and socially just measures. But government action may result from a major climate crisis. In this case, the response is likely to be repressive and extremely inegalitarian. Hurricane Katrina and its aftermath illustrates the kind of immediate response that could be expected.[43] Subsequent policies to cut down on GHG emissions would probably depend on pricing mechanisms: that is, the rich can continue their polluting consumption patterns, while the access of working class people to mobility, warmth and other comforts will be limited.

We are seeing this already, with road pricing, the congestion charge, airline fuel taxes, energy price increases. Many Greens – and some socialists – support fiscal measures, perceiving them as the only effective way to cut down on emissions. Some may even advocate rebates to counteract the effects of energy taxes on the poorest people. This however would partially defeat the object of cutting down on fossil fuel use.

When emergency measures are required, they should be applied fairly, with an emphasis on equality and under workers' and community control: in other words, they should be regulatory, not fiscal. Such measures would also point towards the need for planning – under workers' and community control – to counteract climate change.

Building a movement against climate change

For now, we are faced with small campaigning groups that have a high degree of understanding of the issue, and left political parties which are at best intermittently conscious of the importance of climate change. One task for socialists who understand the issues is to explain the potential (indeed the urgent necessity) for radical social change in the struggle to combat global warming. Capitalism is by its very nature incapable of producing for use value rather than profit; incapable of planning reductions in energy use large enough to solve the problems the planet faces.

A second important issue – highlighted by Katrina – is the crucial importance of social justice. Campaigns should highlight the obscene conspicuous consumption of the rich. They may be small in number, they may not contribute such a lot to emissions, but they act as models for the consumerist society and others are led to aspire to their levels of consumption. No campaign can be credible if it advocates measures like making working class people use their cars less if it does not also do the same for the rich with their limousines and private planes.

As suggested above, campaigns should concentrate on regulatory measures to control GHG emissions, applied equally across society. It should set as its aim a reduction of GHG emissions by Britain of 60 percent within the next ten years and call for the government to fight for an international agreement to get such a reduction on a world scale. We have little time to lose.

A nuclear solution to global warming?

Alice Cutler[44]

Despite the promise to make 2005 the year that politicians would face up to the challenges of climate change, the topic remained a low priority in the election campaign. Since the election, Blair has planned for a return to 'clean' nuclear power.

While plans continue for new roads, airport expansions and business as usual, there seems to be a deep-seated denial in our society about the seriousness of the situation we face. The G8 countries account for 12 percent of the world's population and 62 percent of the total greenhouse gas emissions. It is crucial that we cut our carbon emissions by 60-90 percent to avert the catastrophic consequences of climate change, (IPCC). One sign that people are beginning to acknowledge something has to be done is the resurgence of the nuclear power option, once again being hailed as the clean, safe, carbon-neutral way to produce electricity. James Lovelock, author of *Gaia*, famously came out in favour of the nuclear option to reduce the threat of global warming caused by the burning of fossil fuels. Nuclear power has become a lesser evil in the minds of many and a Government white paper recommending that ten new nuclear stations be built is apparently waiting in the wings for after the election.

Campaigns, such as CND, have traditionally focused on nuclear weapons and on the still unsolved issue of highly dangerous nuclear waste and the health risks from the radiation associated to it. This article is more concerned with nuclear power and to expose the myth that the nuclear option could provide a solution to cutting greenhouse gases.

With the concerns over climate change and in the desperate search to find non-fossil fuel energy sources, the facts about nuclear power have become blurred. Let us look at what the real science of the matter shows.

MYTH: Nuclear power does not create carbon dioxide (CO_2) emissions.
REALITY: Although most reactors do not produce CO_2, the nuclear fuel cycle does.

Uranium mining, milling, processing and enrichment, dealing with nuclear waste and transportation are all carbon intensive processes. The amount of CO_2 produced depends on the grade of uranium ore and the method of enrichment used to process it. The fissile material (U-235) in natural uranium only constitutes around 0.7 per cent, too low for nuclear reactions to occur. This fissile material must be enriched approximately fourfold for it to be able to be used in a reactor. The enrichment process requires enormous processing plants and is hugely energy intensive.

Using favourable assumptions (high-grade, easy-to-process shale ore, diffusion enrichment and an easy method of waste disposal), the nuclear option is estimated to produce as much as one third of the CO_2 produced by gas-fired power stations, per kilowatt hour (kWh) of electricity. Using more pessimistic assumptions, (low-grade ore, hard-to-process granite ore, centrifuge enrichment and difficult waste disposal) nuclear cycle carbon emissions could equal or even exceed those of gas-fired power stations. There is a deficit of research relating to these statistics with no government that uses nuclear power researching the optimistic claims about carbon emissions made by industry funded scientists.

MYTH: There is a limitless supply of fissile material.
REALITY: The U concentration of uranium ore is the most crucial factor in determining carbon emissions and so it is important to note that, as with oil, we have already 'picked the low-hanging fruit'. Most easily obtainable U-rich seams have already been mined and the search for high-grade uranium ore is becoming more difficult. This means that the CO_2 needed to extract equivalent amounts of uranium will inevitably increase in future years. Although very large amounts of uranium exist in the earth's crust and in the sea, the concentrations are so low that they are not viable, as the energy required for extraction would exceed the energy produced. Uranium mining is also a destructive, energy-intensive process that has had disastrous effects on nearby communities, such as those near the former uranium mines in Germany, the Czech Republic, Australia and Canada.

MYTH: Producing UK electricity with nuclear is a good way to cut greenhouse gas emissions.

REALITY: Electricity generation is only responsible for around 25 percent of annual CO_2 production in the UK. The rest is from transportation and domestic, industrial and commercial heating. The electricity we require varies enormously during the day, with high peaks in the morning and evening. However nuclear reactors cannot be switched on/off or cranked up to meet these varying electricity loads: for safety reasons they work at a fixed rate twenty-four hours a day. This means nuclear power could only make a maximum contribution of 25 percent to UK electricity generation. So, the actual potential for CO_2 reduction is only about 5 percent. And of course there are other greenhouse gases as well as carbon dioxide. The end result is that nuclear is a very poor way to reduce UK greenhouse gases.

MYTH: Nuclear power is a cost-effective way to reduce emissions.
REALITY: To build the new reactors being proposed by the nuclear industry (the so-called AP 1000MW reactor) would cost approximately £1.4 - 2 billion per reactor, assuming that ten were built to get economies of scale. The total cost would be around £14 to 20 billion. This would only be possible with massive government subsidies, which would probably be illegal under EU competition rules. A number of studies estimate that nuclear is five to seven times less cost-effective than energy efficiency or renewable energies in reducing CO_2 emissions. The priority should be end-use efficiency: that is measures introduced at the point of electricity consumption.

FACT: The DTI has consistently invested two to three times more in nuclear energy than in renewable and novel sources of energy. In 2004, the figures were £57.8 million on nuclear technologies and only £19 million on renewable sources.[45]

FACT: The estimated time needed to observe legal procedures, carry out public inquiries, training and construction for a nuclear reactor is ten to fifteen years from the decision. We must act sooner than this.

FACT: The Royal Commission on Environmental Pollution (Flowers Report, 1976) said that until a method to deal with nuclear waste has been found, no programme of nuclear fission should be carried out.

40

To this day, no method exists. We should not be embarking on a new nuclear programme without having solved the huge nuclear waste problems of the first programme.

Conclusion

Nuclear power is not a cost-effective, viable or safe solution and it does not address the core problem of unsustainable energy use. The question remains then, why is the Government considering an unpopular return to nuclear power construction? British Energy (which runs most of the UK's nuclear stations) is already supported by the Government to the tune of £300 million a year.

Lord Falconer, head of the Government's legal administration and close colleague of Blair, was, in the early 1990s, chief legal executive for British Nuclear Fuels. There he was instrumental in bringing legal injunctions against Greenpeace anti-nuclear campaigners and seeking sequestrations of Greenpeace assets. BNFL is an important UK company with major US Westinghouse holdings and the Government wants to see returns on its investments in the company. Nuclear technology is a major potential export for the UK; countries such as South Korea, China and Taiwan are all possible customers. Could this be 'the tackling of climate change' that Blair meant when he announced his three-point plan to the UK business leaders back in September 2004? He said then that tackling climate change did not have to be an unbearable burden to business, that the UK could benefit from its leading role in the technology and science, and that we had to deal with rising emissions from rapidly developing countries such as China.

If Blair and the G8 truly wanted to face up to climate change they would reverse their neo-liberal, market system that puts profit above all else and address the dramatically unsustainable energy consumption of the rich and powerful. A return to nuclear power raises the false and dangerous prospect that 'something is being done', that energy remains plentiful and that no radical changes in our lives and consumption patterns are necessary. Meanwhile power literally remains in the hands of massive energy corporations and undermines the development of truly sustainable, community owned, small

41

scale, local, renewable alternatives. It is typical of a viewpoint that seeks to use ever more technology rather than tackle the underlying causes. The nuclear option represents a further example of exploiting a technology for the short-term benefit of the few with complete disregard for future generations or the health of the planet. The anti-nuclear movement in the UK must rise from the ashes and expose the dangerous myths of nuclear power as part of the solution to global warming and instead demand real climate justice.

Nuclear juggernaut moving into top gear

Phil Ward

Recent rejections by government of planning applications for wind farms in Cumbria, along with yet another energy review have led many to believe that New Labour, faced with soaring energy costs and a need to be seen to respond to global warming, will take the nuclear option.

In late February and early March 2006, climate change again came to the top of the media agenda, but this time with a new twist. There was the usual litany of bad news about the extent of the problem. This was surveyed briefly above (see "A large scale geophysical experiment?"), but it is worth mentioning a couple of additional items.

Several recent discoveries show that warming has real, measurable effects: glaciers in Greenland are slipping into the sea at a rate that doubled between 1996 and 2000 and the Antarctic ice cap, which holds 70 percent of the world's water, is now losing water at the same rate as Greenland. This last observation was not predicted: climate models had envisaged a build-up of ice on this continent. Finally, the report in January that north Atlantic sea currents had declined in volume by 30 percent suggests that Europe may experience major weather changes that could harm agriculture.

A leak from the forthcoming IPCC[46] report says that this body is now putting no cap on its predictions of warming this century, apparently because changes are happening so rapidly that it is very difficult to model 'feedback', the phenomenon where temperature rises cause changes that themselves release further greenhouse gases (GHGs) or lead to greater warming. An example is the melting of Siberian permafrost that releases methane – a stronger GHG than carbon dioxide – and also leads to a landscape that absorbs more solar radiation.

The day after the IPCC report, Blair met campaigners from Stop Climate Chaos (SCC)[47] a new UK coalition of charities and NGOs. This has been dubbed as global warming's equivalent to Make Poverty

History. SCC calls for social and economic justice and for countries to act on their own if new international agreements on GHG emission targets cannot be made. In a letter to SCC,[48] Blair rubbishes any idea of a pioneering effort on emissions: 'Action on climate change without America, India and China in agreement, and all committed to a framework for reducing green house gas emissions, is not going to work. That is the brutal truth.'

Meanwhile, Bush, in his state of the union address, promised a lower 'dependence on oil', and promotion of 'renewable alternatives'. Behind this, however, are geopolitical considerations – the need to import less Middle East oil, and advocating nuclear power. Thus, according to one commentator, Bush promotes 'expanded use of nuclear power and so-called 'clean coal' while simultaneously cutting funds for wind, solar, geothermal, hydropower, and energy efficiency programs'.

The cynical agreement for the US to supply nuclear power technology to India fits this scenario neatly. One by-product could be to revitalise the ailing US nuclear power sector (no plant has been ordered since the Three Mile Island explosion in 1979 and the last completed plant has now been operating for 21 years). Assurances that the Indian civil and nuclear sectors would now be kept separate are false, as they are only planning to allow 14 of the 21 reactors (seven are coming on line in the next two years) to be inspected. Their fast-breeder reactors, which can generate nuclear fuel, as well as plutonium for bombs, are designated as military facilities.

Meanwhile, the 2006 UK Energy Review[49] is predicted to overturn the decision in 2003 not to promote nuclear power as a means to deal with GHG emissions. According to Blair, the US 'has a point' when it refuses to exclude nuclear power as an option to reduce GHG emissions. Government research has already been commissioned into new reactor designs.

These moves are complemented in other countries. China has announced it will build 28 new nuclear power stations by 2020. There are over 400 nuclear power stations, operating in over 30 countries and producing 16 percent of the world's electricity[50] (only two percent of total world energy use) which the IAEA[51] advocates increasing to 27 percent by 2030.

Most of the political and safety arguments about nuclear power have been well covered over the 30 years of the movement opposing this technology. What also needs addressing is its effectiveness in combating global warming. The key difficulties with nuclear power were brought up by Alice Cutler in an article in *Socialist Outlook* 6 (May 2005):

- You may be able to generate electricity using nuclear power, but electricity accounts for only 2 percent of world energy usage (the rest is fossil fuels, biomass and hydroelectricity). The inability of nuclear power to deal with this problem of scale is shown by the fact that China's 28 new plants will account for a mere 4 percent of the country's electricity needs in 2020.
- The alternatives to nuclear power are more cost-effective – especially energy conservation.
- The preferred technology for the future, the Pebble Bed Modular Reactor, of which there is a prototype currently undergoing construction in South Africa,[52] poses several new safety issues. It also is likely to produce larger volumes of waste. Worryingly, the developers consider that the system is so safe that costs can be reduced by building the reactors close to towns and without containment, and that they can have lower staffing levels than current designs.[53]
- The richest sources of uranium are becoming exhausted and the enrichment process is very energy-intensive. For these reasons, the carbon dioxide emissions reductions that result from using nuclear power may be grossly over-estimated. This is a hotly contested issue[54] and difficult to judge for a lay person. In some respects, it is a secondary matter, especially when the argument put forward is that reliance on any technology to reduce GHGs plays into the capitalists' hands, when the need is for massive social change.

The question then arises of why many governments are now pursuing nuclear power again. The obvious first answer that it will make lots of profits for the nuclear industry, including those building the stations. But an essential part of this is also the need to maintain centralized control of the energy supply industry. Because of its dangers, nuclear power has the additional attraction that the work force can be placed under quasi-police state discipline. Finally, a small number of companies operate in the sector, and these require protection. This applies particularly to BNFL, which undertakes reprocessing for large parts of the nuclear power community.[55]

Nuclear power reveals in stark terms many of the problems of other technologies advocated by capitalist governments as their answer to climate change. Thus, sequestration technologies – separating carbon dioxide from the combustion gases from fossil fuels and pumping this into oil or gas wells or under the sea – will mainly be add-ons to existing power stations. Biomass for cars (for example palm oil, sunflower oil or 'bioethanol') is adding a huge agribusiness to current oil refining and distribution industries. Many renewable electricity technologies will involve massive, environmentally damaging plant, plugged into the national electricity grids.[56] Developing countries, while having the option of seeking alternatives to this approach, are simply aping richer nations instead.

All the energy-realising technologies presented as alternatives to fossil fuels have significant negative environmental consequences, especially if applied on a scale to meet current energy demands – let alone those of the future. There will also be issues of centralisaton and control of these large-scale technologies. The only feasible way to cut GHGs is a massive cut in energy demand,[57] combined with judicious, planned and community-controlled introduction of alternative energy technologies. To do this, while at the same time improving the quality of life of 95 percent of the people on this planet, will ultimately mean a direct challenge to the capitalist system.

Part two: what politics for what ecology?

No Solution Under Capitalism

Jane Kelly and Phil Ward

There is now no doubt that the ecological crisis of the planet has reached a new stage, one which demands urgent solutions without which the lives of millions of people along with many plant and animal species, will be endangered. While the environmentalists and the Green Party have placed these issues at the top of the public agenda, their solutions, set within a capitalist framework, are utopian. Socialists have to engage with these issues and show how only socialist answers can overcome the crisis of ecology faced by the planet.

The main elements of the crisis have been well rehearsed and are most noticeable in climate change and global warming. First the 1980s and then the 1990s were reported to have been the warmest decades on record. In Britain it is likely that 2003 will turn out to have been the warmest year ever recorded. The Greenland ice cap is thinning, leading to rising sea levels and the imperilling of specific islands and coastal regions. Glaciers are retreating, much of the world is heating up, leading to desertification, while other areas are experiencing excessive rainfall. Food production is threatened, especially with the present agricultural practices. Within general global warming, some areas may become much colder: for example, if the Gulf Stream changes course owing to the melting of the Polar ice, Britain and the rest of northern Europe will experience a tundra climate. Most think that greenhouse gases are the cause of all these changes.[58]

The use of fossil fuels for electricity generation, the heating and cooling of domestic and commercial buildings and in transport releases a variety of toxic substances into the air, especially noticeable in cities. This has led to an alarming increase of respiratory illnesses such as asthma, bronchitis and lung cancer. Other industrial pollutants include asbestos in building materials[59] and mercury in batteries.

There is also both water and soil pollution and deterioration. Industrial and household waste is carried off in the waters of the world, reducing them to sewers. Rivers, lakes and now seas are affected by the twin disasters of toxic accumulation and fertiliser build-up. The combination of toxins and the proliferation of algae and water plants are exhausting the oxygen in the water, resulting in a massive loss of aquatic life. Oil spills and the seeping of petroleum from underwater drilling are adding to the pollution alongside chemical and radioactive waste.

Capitalist market pressures have led to the misuse of fertilisers, pesticides, and animal feeds; the practice of monoculture and the replacement only of those minerals needed for particular crops to grow. This is leading to poisonous and depleted soils, which in turn produce mineral-deficient food. Many argue that this is an element in the rise of degenerative diseases. Intensive farming methods have also led to the spread of animal diseases like BSE. Overfishing has led to the near-disappearance of many fish species.

Among the most dramatic manifestations of the ecological crisis is the destruction of the world's forests. In the last fifty years one third of the world's woodland has disappeared. While in the industrialised countries, where many forests disappeared centuries ago, trees are dying from air and soil pollution, in neo-colonial countries deforestation is at the heart of the ecological crisis. Brazil's tropical rain forests are disappearing at an alarming rate, cut down or burnt to create short-term grazing land for cattle to produce quick profits for big landowners. Other areas of tropical forest are cut down for use as timber. Either way this process, exacerbated by huge fires such as those in Indonesia in 1997, is an additional factor in the greenhouse effect and in the destruction of animal and plant species, fifty per cent of which live in tropical forests.

While much of this misuse can be laid at the door of capitalism and imperialism, the primary motive of which is profit, the rulers of the ex-USSR and Eastern Europe were also complicit in ecological pillage. The bureaucratic nature of their central planning meant that their system was as neglectful of the environment as the imperialist countries. Huge projects, such as changing the course of Siberian rivers, are being replicated in China with the Yangtze Diversion Scheme.

Marx and Engels

The workers' movements have been slow to take these issues up, even though in the nineteenth century Engels wrote of the appalling conditions of working class slums. He describes a stream in Bolton thus: 'a dark-coloured body of water, which leaves the beholder in doubt whether it is a brook or a long stream of stagnant puddles, flows through the town and contributes its share to the total pollution of the air, by no means pure without it.'[60] He notes that above Dulcie Bridge in Manchester 'are tanneries, bone mills, and gasworks, from which all drains and refuse find their way into the Irk, which receives further the contents of all the neighbouring sewers and privies. It may be easily imagined, therefore, what sort of residue the stream deposits.' [61]

Engels goes on to discuss the houses and cottages, the filth, the pig pens amidst it all – all the result of 'the industrial epoch' and the motivating force of profit,

> ...the value of the land rose with the blossoming out of manufacture, and the more it rose, the more madly was the work of building up carried on, without reference to the health or comfort of the inhabitants, with sole responsibility to the highest possible profit on the principle that **no hole is so bad but that some poor creature must take it who can pay for nothing better.**[62]

The environmental crisis we now face is qualitatively greater than when Marx and Engels were writing. But it is not true that their work ignored these questions: like the accusation of gender blindness in their work, this charge is refuted by reference to their actual writing. Moreover, they located capitalism's central motivating principle of profit – before and beyond any attention to the usefulness of the objects produced or the effects of the process on the environment and on the human producers. This is a key understanding: while capitalism can choose whether to take into account the environmental impact of its production processes, depending on a variety of factors including the opposition of environmental activists and the rate of profit, it will not be able to change its central tenet, the law of value.

It is to their great credit that environmental activists have placed the issue of the unsustainability of capitalism's destructive processes at the centre of the political agenda. This has forced even New Labour to make gestures towards renewable power such as their introduction of offshore wind turbines. They are interested in the market for wind turbines and the UK also has the largest wind resources in Europe, which means the costs are lower here than anywhere else. Furthermore, we are paying the wind power premium through our electricity bills and the government has set itself quite a high target for electricity to be generated by renewables under the renewables obligation.

The Green Party

While campaigns and movements for the environment have put these issues onto the agenda, in Britain it is the Green Party that has systematically taken up the challenge at an electoral level. The importance of the issue is reflected in the two and a quarter million votes they received in the European Election in 1989, and although they have not gained anything like that since, they have a number of local council seats and three seats on the Greater London Assembly. They regularly come fourth in elections after the three main parties. In raising the right issues, they attract many concerned with the crisis including young people. In several cases, Socialist Alliance candidates have come close to the votes of the Green Party, even ahead of them occasionally, but there has been a division of labour, for the Socialist Alliance tends not to take up green issues adequately, and the Green Party does not provide socialist solutions.

The Green Party itself is criss-crossed by many different types of politics from socialist to right wing ideas. Many of their spokespeople have very good positions on a variety of issues beyond the environment. For example Caroline Lucas, one of their MEPs, speaks on Stop the War platforms, works with the anti-globalisation protesters, is involved in the European Social Forums and is opposed to the current asylum legislation. But a quick reference to the Green Party's positions on the alternative to unfettered capitalism reveals a politics that is nothing short of pathetic.

In their anti-globalisation document, they fail to recognise the interrelationship of the bourgeois state and the capitalist system, calling for 'regional blocs, such as the EU and North America' to play a role in countering globalisation. They continue, 'Indeed these two blocs are the only ones politically and economically powerful enough to be a counter-weight to overcome the forces which are the major beneficiaries from globalisation – transnational companies and international capital.' And since the Bush regime is unlikely to comply with this proposal, 'The EU must therefore take on the mantle as the major engine for change.' [63]

Ignoring the weight of the imperialist drive to maintain and increase profits, which is the motor of economic globalisation and its military might, they plaintively call for the re-localisation and democratisation of sustainable economies worldwide. 'The EU should therefore move away from its present emphasis on ruthless internal and external competition leading to unsustainable growth.' [64]

At the heart of their politics is a developed analysis of the environmental crisis combined with an under-theorized and mistaken interpretation of the role of the state. Like bourgeois liberals and social democrats, they see the state as essentially neutral. It may be captured by reactionary elements and so pass bad laws, but conversely it could be taken over by progressive forces and be a power for good. These views rule out any socialist solutions as unnecessary. Recognising the misrule and environmental damage caused in the ex-USSR and beyond, and identifying it as 'socialist' the Greens do not differ fundamentally from social democracy in the belief that capitalism can be reformed.

The left and ecology today

While revolutionary Marxists understand that capitalism cannot be reformed but has to be overthrown, our programme for environmental change is not so well thought through. But as in other matters we recognise that it is necessary to fight now for reforms to ameliorate the situation. To do so, it will be necessary to stand very firmly within a Marxist framework in order to analyse the situation and put forward the necessary demands. Some recent events clearly show the problems we face.

Congestion charging, introduced by Ken Livingstone in central London in 2003, immediately provoked a heated exchange between people who all considered themselves Marxists. On the one side were those who supported the charge as a way of reducing pollution and making travellers use public transport. Some, abandoning a class analysis, argued that the poorest people do not own cars anyhow, so it would only hit the better off. While it may be true that the poorest sections cannot afford their own car, this does not cover the whole working class. Working class people own and need to use cars – for example mothers dropping their children off at nursery and school on their way to work. And as public transport is both badly organised and expensive, it may be cheaper for some people to use their car. Unless public transport fares were at the same time abolished or reduced to a minimum, they argued, we should oppose the charge. The debate on what at first sight seemed a simple matter (to both sides) revealed how little green issues have been discussed except at a very general level amongst socialists.

Taxing our way out of the problem?

Behind the congestion charge issue is a fundamental debate, with the Greens and New Labour on one side and socialists on the other. This is the issue of the use of taxation and other pricing mechanisms to compensate for the 'failure of the market' to reduce or eliminate environmentally damaging practices. Both New Labour (tentatively) and the Greens (with some vigour) advocate this central mechanism. Lying behind it is a reactionary, behaviourist view of 'human nature': that we will inevitably act selfishly and against the 'common good', unless coerced to do otherwise. Taxes and prices then become part of that coercion. The most explicit expression of that view, still quoted nowadays by people who should know better is *The Tragedy of the Commons* by Garrett Hardin (1968).[65]

Hardin's reactionary views on human nature are also seen in the work of many Greens, who seem to have a particularly pessimistic and despairing disposition. They view individuals and their 'consumptionist' behaviour as the cause of environmental problems and at the same time have no understanding of agency in achieving political change. They are thus wedded to their electoralist strategy. [66]

Green socialism

It is, of course, useful to state the principles which should frame the specifics of an environmentalist programme, and also the ways in which socialist and green demands are linked. So the demand that production be for use rather than exchange value immediately throws into question the way in which agricultural goods in the neo-colonial world are tied to export demands, rather than being used to feed the local population. This then highlights indebtedness. In order to change agricultural practices some countries will not only need their debt to be cancelled, but economic support while they transform their agriculture from the production of a single export commodity to as near self-sufficiency as can be achieved.

All this is impossible within a capitalist economy. The drive for profit contradicts the motive of production for need. In addition neo-colonial countries whose systems are often barely democratic, are characterised by what is called combined and uneven development: that is combining developed industrial processes alongside techniques of production and social relations – including the position of women and children – from another mode, even semi-feudal. Trotsky argued that it was impossible for such regimes to achieve even basic bourgeois rights without recourse to a process leading to socialism. The fight for democratic rights would spill over into socialist demands. Thus environmental demands cannot be achieved outside of socialist change.

This should not lead us to put off such demands until after socialism is achieved. Reforms can be fought for now, such as changing from the use of fossil fuels and nuclear power to renewable energy; a huge energy conservation programme and freely available home insulation and new designs for buildings so they need minimal heating and cooling, a transport policy that would genuinely reduce our need to use cars. We need to develop international plans on the use of water, to deal with interstate conflicts over water such as between Iraq, Syria and Turkey; we need to develop a series of measures to deal with the likely increase of areas of both drought and floods.

None of the above can be successfully achieved without the control of ordinary working people; issues of workers' control, workers' democracy and socialist solutions are paramount. But we also have to argue that a democratic socialist society can provide for everyone's needs without destroying the environment. When we talk of abundance this has nothing to do with the waste resulting from commodity production. There are five hundred million cars in the world, each of them stationary for ninety-five per cent of its life; many will be destroyed because of what used to be called 'built-in obsolescence'.[67] These can be replaced by free public transport, shared ownership and changes in the way we organise our lives.

Moreover, the inefficiency of the bourgeois nuclear family, with each household having to do its own washing, cooking, child care, etc. could be overcome by changed living patterns. Domestic labour could be collectivised – there could be neighbourhood restaurants, laundries, local child care facilities – all of which would help to liberate women at the same time as saving on resources. While many ecologists propose beneficial measures like home-based energy saving such as the generation of methane from sewage, or growing your own vegetables, unless these are combined with promoting collective living, women, children and men will become more not less dependent on the nuclear family.

As Hans Magnus Enzensberger wrote in 1974, it is essential for the workers' movement to take action in defence of the environment, for it is also in defence of democracy and the future of humanity:

In reality, capitalism's policy on the environment, raw materials, energy and population, will put an end to the last liberal illusions. That policy cannot even be conceived without increasing repression and regimentation. Fascism has already demonstrated its capabilities as a saviour in extreme crisis situations and as the administrator of poverty. In an atmosphere of panic and uncontrollable emotions – that is to say, in the event of an ecological catastrophe which is directly perceptible on a mass scale – the ruling class will not hesitate to have recourse to such solutions. The ability of the masses to see the connection between the mode of production and the crisis in such a situation and to react offensively cannot be assumed. It depends on the degree of politicisation and organisation achieved by then.[68]

Organizing ecological revolution

John Bellamy Foster

My subject – organizing ecological revolution – has as its initial premise that we are in the midst of a global environmental crisis of such enormity that the web of life of the entire planet is threatened and with it the future of civilization.[69]

This is no longer a very controversial proposition. To be sure, there are different perceptions about the extent of the challenge that this raises. At one extreme there are those who believe that since these are human problems arising from human causes they are easily solvable. All we need are ingenuity and the will to act. At the other extreme there are those who believe that the world ecology is deteriorating on a scale and with a rapidity beyond our means to control, giving rise to the gloomiest forebodings.

Although often seen as polar opposites these views nonetheless share a common basis. As Paul Sweezy observed they each reflect "the belief that if present trends continue to operate, it is only a matter of time until the human species irredeemably fouls its own nest" (*Monthly Review*, June 1989).

The more we learn about current environmental trends the more the unsustainability of our present course is brought home to us. Among the warning signs:

- There is now a virtual certainty that the critical threshold of a 2°C (3.6°F) increase in average world temperature above the preindustrial level will soon be crossed due to the buildup of greenhouse gases in the atmosphere. Scientists believe that climate change at this level will have portentous implications for the world's ecosystems. The question is no longer whether significant climate change will occur but how great it will be (International Climate Change Task Force, *Meeting the Climate Challenge*, January 2005, www.americanprogress.org).
- There are growing worries in the scientific community that the estimates of the rate of global warming provided by the United

Nations Intergovernmental Panel on Climate Change (IPCC), which in its worst case scenario projected increases in average global temperature of up to 5.8° C (10.4° F) by 2100, may prove to be too low. For example, results from the world's largest climate modelling experiment based in Oxford University in Britain indicate that global warming could increase almost twice as fast as the IPCC has estimated (*London Times*, January 27, 2005).

- Experiments at the International Rice Institute and elsewhere have led scientists to conclude that with each 1°C (1.8°F) increase in temperature, rice, wheat, and corn yields could drop 10 percent (Proceedings of the National Academy of Sciences, July 6, 2004; Lester Brown, *Outgrowing the Earth*).

- It is now clear that the world is within a few years of its peak oil production (known as Hubbert's Peak). The world economy is therefore confronting diminishing and ever more difficult to obtain oil supplies, despite a rapidly increasing demand (Ken Deffeyes, *Hubbert's Peak*; David Goodstein, *Out of Gas*). All of this points to a growing world energy crisis and mounting resource wars.

- The planet is facing global water shortages due to the drawing down of irreplaceable aquifers, which make up the bulk of the world's fresh water supplies. This poses a threat to global agriculture, which has become a bubble economy based on the unsustainable exploitation of ground water. One in four people in the world today do not have access to safe water (Bill McKibben, *New York Review of Books*, September 25, 2003).

- Two thirds of the world's major fish stocks are currently being fished at or above their capacity. Over the last half-century 90 percent of large predatory fish in the world's oceans have been eliminated (Worldwatch,, *Vital Signs* 2005).

- The species extinction rate is the highest in 65 million years with the prospect of cascading extinctions as the last remnants of intact ecosystems are removed. Already the extinction rate is approaching 1,000 times the "benchmark" or natural rate (*Scientific American*, September 2005). Scientists have pinpointed twenty-five hot spots on land that account for 44 percent of all vascular plant species and 35 percent of all species in four vertebrate groups, while taking up only 1.4 percent of the world's land surface. All of these hot spots are now threatened with rapid annihilation due to human causes (*Nature*, February 24, 2000).

- According to a study published by the National Academy of Sciences in 2002, the world economy exceeded the earth's regenerative capacity in 1980 and by 1999 had gone beyond it by as much as 20 percent. This means, according to the study's authors, that "it would require 1.2 earths, or one earth for 1.2 years, to regenerate what humanity used in 1999" (Matthis Wackernagel, et. al, "Tracking the Ecological Overshoot of the Human Economy," *Proceedings of the National Academy of Sciences*, July 9, 2002).
- The question of the ecological collapse of past civilizations from Easter Island to the Mayans is now increasingly seen as extending to today's world capitalist system. This view, long held by environmentalists, has recently been popularized by Jared Diamond in his book *Collapse*.

These and other warning bells indicate that the present human relation to the environment is no longer supportable. The most developed capitalist countries have the largest per capita ecological footprints, demonstrating that the entire course of world capitalist development at present represents a dead end.

The main response of the ruling capitalist class when confronted with the growing environmental challenge is to "fiddle while Rome burns." To the extent that it has a strategy, it is to rely on revolutionizing the forces of production, i.e., on technical change, while keeping the existing system of social relations intact. It was Karl Marx who first pointed in *The Communist Manifesto* to "the constant revolutionizing of production" as a distinguishing feature of capitalist society. Today's vested interests are counting on this built-in process of revolutionary technological change coupled with the proverbial magic of the market to solve the environmental problem when and where this becomes necessary.

In stark contrast, many environmentalists now believe that technological revolution alone will be insufficient to solve the problem and that a more far-reaching social revolution aimed at transforming the present mode of production is required.

Historically addressing this question of the ecological

transformation of society means that we need to ascertain: (1) where the world capitalist system is heading at present, (2) the extent to which it can alter its course by technological or other means in response to today's converging ecological and social crises, and (3) the historical alternatives to the existing system. The most ambitious attempt thus far to carry out such a broad assessment has come from the Global Scenario Group (www.gsg.org), a project launched in 1995 by the Stockholm Environmental Institute to examine the transition to global sustainability. The Global Scenario Group has issued three reports – *Branch Points* (1997), *Bending the Curve* (1998), and their culminating study, *Great Transition* (2002). In what follows I will focus on the last of these reports, the *Great Transition*.[70]

As its name suggests, the Global Scenario Group employs alternative scenarios to explore possible paths that society caught in a crisis of ecological sustainability might take. Their culminating report presents three classes of scenarios: Conventional Worlds, Barbarization, and Great Transitions. Each of these contains two variants. Conventional Worlds consists of Market Forces and Policy Reform. Barbarization manifests itself in the forms of Breakdown and Fortress World. Great Transitions is broken down into Eco-communalism and the New Sustainability Paradigm. Each scenario is associated with different thinkers: Market Forces with Adam Smith; Policy Reform with John Maynard Keynes and the authors of the 1987 Brundtland Commission report; Breakdown with Thomas Malthus; Fortress World with Thomas Hobbes; Eco-communalism with William Morris, Mahatma Gandhi, and E. F. Schumacher; and the New Sustainability Paradigm with John Stuart Mill.

Within the Conventional Worlds scenarios Market Forces stands for naked capitalism or neoliberalism. It represents, in the words of the *Great Transition* report, "the firestorm of capitalist expansion." Market Forces is an unfettered capitalist world order geared to the accumulation of capital and rapid economic growth without regard to social or ecological costs. The principal problem raised by this scenario is its rapacious relation to humanity and the earth.

The drive to amass capital that is central to a Market Forces regime is best captured by Marx's general formula of capital (though not referred to in the *Great Transition* report itself). In a society of simple commodity production (an abstract conception referring to pre-capitalist economic formations in which money and the market play a subsidiary role), the circuit of commodities and money exists in a form, C–M–C, in which distinct commodities or use values constitute the end points of the economic process. A commodity C embodying a definite use value is sold for money M which is used to purchase a different commodity C. Each such circuit is completed with the consumption of a use value.

In the case of capitalism, or generalized commodity production, however, the circuit of money and commodities begins and ends with money, or M–C–M. Moreover, since money is merely a quantitative relationship such an exchange would have no meaning if the same amount of money were acquired at the end of the process as exchanged in the beginning, so the general formula for capital in reality takes the form of M–C–M′, where M′ equals M + m or surplus value.[71] What stands out, when contrasted with simple commodity production, is that there is no real end to the process, since the object is not final use but the accumulation of surplus value or capital. M–C–M′ in one year therefore results in the m being reinvested, leading to M–C–M″ in the next year and M–C–M‴ the year after that, ad infinitum. In other words, capital by its nature is self-expanding value.

The motor force behind this drive to accumulation is competition. The competitive struggle ensures that each capital or firm must grow and hence must reinvest its "earnings" in order to survive.

Such a system tends toward exponential growth punctuated by crises or temporary interruptions in the accumulation process. The pressures placed on the natural environment are immense and will lessen only with the weakening and cessation of capitalism itself. During the last half-century the world economy has grown more than seven-fold while the biosphere's capacity to support such expansion has if anything diminished due to human ecological depredations (Lester Brown, *Outgrowing the Earth*).

The main assumption of those who advocate a Market Forces solution to the environmental problem is that it will lead to increasing efficiency in the consumption of environmental inputs by means of technological revolution and continual market adjustments. Use of energy, water, and other natural resources will decrease per unit of economic output. This is often referred to as 'dematerialization'. However, the central implication of this argument is false. Dematerialization, to the extent that it can be said to exist, has been shown to be a much weaker tendency than M–C–M'. As the *Great Transition* report puts it, "The 'growth effect' outpaces the 'efficiency effect.'"

This can be understood concretely in terms of what has been called the Jevons Paradox, named after William Stanley Jevons who published *The Coal Question* in 1865. Jevons, one of the founders of neoclassical economics, explained that improvements in steam engines that decreased the use of coal per unit of output also served to increase the scale of production as more and bigger factories were built. Hence, increased efficiency in the use of coal had the paradoxical effect of expanding aggregate coal consumption.

The perils of the Market Forces model are clearly visible in the environmental depredations during the two centuries since the advent of industrial capitalism, and especially in the last half-century. "Rather than abating" under a Market Forces regime, the *Great Transition* report declares, "the unsustainable process of environmental degradation that we observe in today's world would [continue to] intensify. The danger of crossing critical thresholds in global systems would increase, triggering events that would radically transform the planet's climate and ecosystems." Although it is "the tacit ideology" of most international institutions, Market Forces leads inexorably to ecological and social disaster and even collapse. The continuation of "'business-as-usual' is a utopian fantasy."

A far more rational basis for hope, the report contends, is found in the Policy Reform scenario. "The essence of the scenario is the emergence of the political will for gradually bending the curve of development toward a comprehensive set of sustainability targets," including peace, human rights, economic development; and environmental quality. This is essentially the Global Keynesian strategy advocated by the Brundtland Commission Report in the late 1980s—an expansion of the welfare state, now conceived as an environmental welfare state, to the entire world. It represents the promise of what environmental sociologists call "ecological modernization."

The Policy Reform approach is prefigured in various international agreements such as the Kyoto Protocol on global warming and the environmental reform measures advanced by the earth summits in Rio in 1992 and Johannesburg in 2002. Policy Reform would seek to decrease world inequality and poverty through foreign aid programs emanating from the rich countries and international institutions. It would promote environmental best practices through state-induced market incentives. Yet, despite the potential for limited ecological modernization, the realities of capitalism, the *Great Transition* report contends, would collide with Policy Reform. This is because Policy Reform remains a Conventional Worlds scenario—one in which the underlying values, lifestyles, and structures of the capitalist system endure. "The logic of sustainability and the logic of the global market are in tension. The correlation between the accumulation of wealth and the concentration of power erodes the political basis for a transition." Under these circumstances the "lure of the God of Mammon and the Almighty dollar" will prevail.

The failure of both of the Conventional Worlds scenarios to alleviate the problem of ecological decline means that Barbarization threatens: either Breakdown or the Fortress World. Breakdown is self-explanatory and to be avoided at all costs. The Fortress World emerges when "powerful regional and international actors comprehend the perilous forces leading to Breakdown" and are able to guard their own interests sufficiently to create "protected enclaves." Fortress World is a planetary apartheid system, gated and maintained by force, in which the gap between global rich and global poor constantly widens and the differential access to environmental resources and amenities increases sharply. It consists of "bubbles of privilege amidst oceans of misery.... The elite[s] have halted barbarism at their gates and enforced a kind of environmental management and uneasy stability." The general state of the planetary environment, however, would continue to deteriorate in this scenario leading either to a complete ecological Breakdown or to the achievement through revolutionary struggle of a more egalitarian society, such as Eco-communalism.

This description of the Fortress World is remarkably similar to the scenario released in the 2003 Pentagon report, "Abrupt Climate Change and its Implications for United States National Security" (see "The Pentagon and Climate Change,"[72] *Monthly Review*, May 2004).

The Pentagon report envisioned a possible shutdown due to global warming of the thermohaline circulation warming the North Atlantic, throwing Europe and North America into Siberia-like conditions. Under such unlikely but plausible circumstances relatively well-off populations, including those in the United States, are pictured as building "defensive fortresses" around themselves to keep masses of would-be immigrants out. Military confrontations over scarce resources intensify.

Arguably naked capitalism and resource wars are already propelling the world in this direction at present, though without a cause as immediately earth-shaking as abrupt climate change. With the advent of the War of Terror, unleashed by the United States against one country after another since September 11, 2001, an "Empire of Barbarism"[73] is making its presence felt (*Monthly Review*, December 2004).

Still, from the standpoint of the Global Scenario Group, the Barbarization scenarios are there simply to warn us of the worst possible dangers of ecological and social decline. A Great Transition, it is argued, is necessary if Barbarization is to be avoided.

Theoretically, there are two Great Transitions scenarios envisioned by the Global Scenario Group: Eco-communalism and the New Sustainability Paradigm. Yet Eco-communalism is never discussed in any detail, on the grounds that for this kind of transformation to come about it would be necessary for world society first to pass through Barbarization. The social revolution of Eco-communalism is seen, by the Global Scenario Group authors, as lying on the other side of Jack London's *Iron Heel*. The discussion of Great Transition is thus confined to the New Sustainability Paradigm.

The essence of the New Sustainability Paradigm is that of a radical ecological transformation that goes against unbridled "capitalist hegemony" but stops short of full social revolution. It is to be carried out primarily through changes in values and lifestyles rather than transformation of social structures. Advances in environmental technology and policy that began with the Policy Reform scenario, but that were unable to propel sufficient environmental change due to the dominance of acquisitive norms, are here supplemented by a "lifestyle wedge."

In the explicitly utopian scenario of the New Sustainability

Paradigm the United Nations is transformed into the "World Union," a true "global federation." Globalization has become "civilized." The world market is fully integrated and harnessed for equality and sustainability not just wealth generation. The War on Terrorism has resulted in the defeat of the terrorists. Civil society, represented by non-governmental organizations (NGOs), plays a leading role in society at both the national and global levels. Voting is electronic. Poverty is eradicated. Typical inequality has decreased to a 2–3:1 gap between the top 20 percent and bottom 20 percent of society. Dematerialization is real, as is the polluter pays principle. Advertising is nowhere to be seen. There has been a transition to a solar economy. The long commute from where people live to where they work is a thing of the past; instead there are "integrated settlements" that place home, work, retail shops, and leisure outlets in close proximity to each other. The giant corporations have become forward-looking societal organizations, rather than simply private entities. They are no longer concerned exclusively with the economic bottom line but have revised this "to include social equity and environmental sustainability, not only as a means to profit, but as ends."

Four agents of change are said to have combined to bring all of this about: (1) giant transnational corporations, (2) intergovernmental organizations such as the United Nations, World Bank, International Monetary Fund, and World Trade Organization, (3) civil society acting through NGOs, and (4) a globally aware, environmentally-conscious, democratically organized world population.

Underpinning this economically is the notion of a stationary state, as depicted by Mill in his *Principles of Political Economy* (1848) and advanced today by the ecological economist Herman Daly. Most classical economists—including Adam Smith, David Ricardo, Thomas Malthus, and Karl Marx—saw the spectre of a stationary state as presaging the demise of the bourgeois political economy. In contrast, Mill, who Marx (in the afterword to the second German edition of *Capital*) accused of a "shallow syncretism," saw the stationary state as somehow compatible with existing productive relations, requiring only changes in distribution. In the New Sustainability Paradigm scenario, which takes Mill's view of the stationary state as its inspiration, the basic institutions of capitalism remain intact, as do the

64

fundamental relations of power, but a shift in lifestyle and consumer orientation mean that the economy is no longer geared to economic growth and the enlargement of profits, but to efficiency, equity, and qualitative improvements in life. A capitalist society formerly driven to expanded reproduction through investment of surplus product (or surplus value) has been replaced with a system of simple reproduction (Mill's stationary state), in which the surplus is consumed rather than invested. The vision is one of a cultural revolution supplementing technological revolution, radically changing the ecological and social landscape of capitalist society, without fundamentally altering the productive, property, and power relations that define the system.

In my view, there are both logical and historical problems with this projection. It combines the weakest elements of utopian thinking (weaving a future out of mere hopes and wishes—see Bertell Ollman, "The Utopian Vision of the Future," *Monthly Review*, July-August 2005) with a "practical" desire to avoid a sharp break with the existing system. The failure of the Global Scenario Group to address its own scenario of Eco-communalism is part and parcel of this perspective, which seeks to elude the question of the more thoroughgoing social transformation that a genuine Great Transition would require.

The result is a vision of the future that is contradictory to an extreme. Private corporations are institutions with one and only one purpose: the pursuit of profit. The idea of turning them to entirely different and opposing social ends is reminiscent of the long-abandoned notions of the "soulful corporation" that emerged for a short time in the 1950s and then vanished in the harsh light of reality. Many changes associated with the New Sustainability Paradigm would require a class revolution to bring about. Yet, this is excluded from the scenario itself. Instead the Global Scenario Group authors engage in a kind of magical thinking – denying that fundamental changes in the relations of production must accompany (and sometimes even precede) changes in values. No less than in the case of the Policy Reform Scenario—as pointed out in The Great Transition report itself – the "God of Mammon" will inevitably overwhelm a value-based Great Transition that seeks to escape the challenge of the revolutionary transformation of the whole society.

Put simply, my argument is that a global ecological revolution worthy of the name can only occur as part of a larger social – and I would insist, socialist – revolution. Such a revolution, were it to generate the conditions of equality, sustainability, and human freedom worthy of a genuine Great Transition, would necessarily draw its major impetus from the struggles of working populations and communities at the bottom of the global capitalist hierarchy. It would demand, as Marx insisted, that the associated producers rationally regulate the human metabolic relation with nature. It would see wealth and human development in radically different terms than capitalist society.

In conceiving such a social and ecological revolution, we can derive inspiration, as Marx did, from the ancient Epicurean concept of "natural wealth."[74] As Epicurus observed in his Principal Doctrines, "Natural wealth is both limited and easily obtainable; the riches of idle fancies go on forever." It is the unnatural, unlimited character of such alienated wealth that is the problem. Similarly, in what has become known as the Vatican Sayings, Epicurus stated: "When measured by the natural purpose of life, poverty is great wealth; limitless wealth is great poverty." Free human development, arising in a climate of natural limitation and sustainability is the true basis of wealth, of a rich, many-sided existence; the unbounded, pursuit of wealth is the primary source of human impoverishment and suffering. Needless to say, such a concern with natural well-being, as opposed to artificial needs and stimulants, is the antithesis of capitalist society and the precondition of a sustainable human community.

A Great Transition therefore must have the characteristics implied by the Global Scenario Group's neglected scenario: Eco-communalism. It must take its inspiration from William Morris, one of the most original and ecological followers of Karl Marx, from Gandhi, and from other radical, revolutionary and materialist figures, including Marx himself, stretching as far back as Epicurus. The goal must be the creation of sustainable communities geared to the development of human needs and powers, removed from the all-consuming drive to accumulate wealth (capital).

As Marx wrote, the new system "starts with the self-government of the communities" (Marx and Engels, *Collected Works*, vol. 24, p. 519; Paul Burkett, "Marx's Vision of Sustainable Human Development"). The creation of an ecological civilization requires a social revolution; one that, as Roy Morrison explains, needs to be organized democratically from below: "community by community... region by region" (Ecological Democracy). It must put the provision of basic human needs—clean air, unpolluted water, safe food, adequate sanitation, social transport, and universal health care and education, all of which require a sustainable relation to the earth – ahead of all other needs and wants. Such a revolutionary turn in human affairs may seem improbable. But the continuation of the present capitalist system for any length of time will prove impossible – if human civilization and the web of life as we know it are to be sustained.

Climate change:
the biggest challenge facing humanity

International Socialist Group[75]

The ecological crisis, particularly climate change, is the biggest challenge facing humanity. But the governmental and corporate response has been woefully inadequate. Evidence mounts by the week of the seriousness of the situation. Last year's hurricane season was just another reminder that climate change (or more precisely the destabilisation of the world's climatic system) is happening faster than most predictions. The last decade was the warmest on record, the last year was again the warmest on record. This is having a devastating impact on communities around the world through increased floods, droughts, and extreme weather events.

The melting of the polar icecaps will raise sea levels, wiping out low-lying countries and coastal communities. Low-lying atoll countries are already facing this. Mountain glaciers are thawing across the globe. The melt-rate in Greenland in accelerating the most recent study reduces its projected melt time from 1,000 to between 100 and 300 years. This would raise sea levels by seven metres.

Deserts are expanding and agricultural land is being destroyed. Water is becoming more scarce and it will increasingly become the cause of conflict. Millions will be driven from where they live to become climate change migrants attempting to seek refuge in other countries. The pace of these events is increasing and they will continue to happen unless effective international action is taken to halt the process. It will be the poor and the dispossessed of the world who will pay the highest price.

Dramatic reduction in emissions

The only solution to this problem is a dramatic reduction in greenhouse gas emissions, in particular carbon dioxide. Most studies suggest that a reduction of at least 60 percent of current carbon dioxide emissions by 2050 is necessary to reverse the damage now being done. Whilst renewable energy sources can play an important part in this, alternatives alone cannot meet such a target. A major reduction in the fossil fuel burn is unavoidable. Even then it could take a hundred years for the climate system to stabilise. Our job must be to ensure that fossil fuel burn is reduced and that it is not the poor who carry the burden of this crisis.

Real change can only come from government action, which has to provide the framework for such changes in life style to take place. But tackling the crisis will involve radical lifestyle changes, and individuals have an important role to play in this. Car usage, as it has developed over the past hundred years in an individualistic society, is completely unsustainable, as is the expansion of air travel at its current rate. The number of air passengers is set to escalate over the next ten years whilst aviation fuel remains uniquely free of tax.

The throwaway economies of the western world generated by the logic of the market are an environmental disaster. The way living space is used in the rich countries of the world cannot continue. The systematic corporate abuse of the environment, which takes place right across the world is unacceptable – as is the current globalised nature of food production which transports food unnecessarily over thousands of miles.

Tackling all this will involve major changes in the way society is structured. It involves major changes in transport systems with a big reduction in private car usage and in air travel. It involves a radical redesign of towns and cities to reduce energy usage and unnecessary travel. It means the local production of food and a major reduction in food miles and a challenge to the role of the supermarkets. It means a more efficient use of living space to restrict energy consumption. All this relies on decisive governmental action.

Emerging nations

Meanwhile economic development, in particular the emergence of China and India as major economic powers, means that far from carbon emissions being reduced they are about to rise to new levels.

China, which has no oil but large coal reserves, is said to be bringing into commission a new coal-fired power station every two weeks, and it is building forty new nuclear power plants. And with car usage escalating each year China is searching the oil markets for long-term supplies. Emerging nations such as China are unlikely to accept limitations on their growth and development whilst the rich countries of the West remain unprepared to make significant changes.

If third world countries are to avoid this crisis they have to avoid the free-for-all model of globalised capitalism that is on offer. It is in the dependant countries of Asia and Africa and Latin America that the relationship between economic conditions and the ecological crisis becomes most clear. There, as was noted in the World Congress text,

> *Over 800 million people are malnourished, 40 million die every year from hunger or diseases caused by malnutrition. Almost two billion do not have regular access to clean drinking water; 25 million die as a result every year. One and a half billion human beings suffer from an acute lack of firewood, their only source of energy. In this part of the world, there is a grave shortage of food, water and fuel, the three essential elements for people's very lives.*

The ability to tackle climate change and the ecological crisis is linked directly to the domination of these countries by the WTO, the IMF, the World Bank and the G8. They cannot tackle climate change effectively whilst saddled with massive debt from western banks and their farmers are made bankrupt by subsidised competition. WTO neo-liberal rules force local markets to open up to international competition under grossly unfair conditions. This worsens the conditions of dependency, undermines social conditions and leads to an irrational increase in international trade even in food products.

International treaty

Climate change is therefore both a political and an international issue. The earth has only one atmosphere. An international solution is the only one that will ultimately count since measures taken by one country can be cancelled out by the actions of another. This is what makes an effective international treaty on climate change, taking into account the conditions faced by third world countries, so important. The December demonstrations rightly demanded, 'that the entire world community move as rapidly as possible to a stronger emissions reductions treaty that will be both equitable and effective in stabilising greenhouse gases and preventing dangerous climate change'. It is a good demand around which to build an international movement.

Kyoto is not the answer. It calls for the stabilisation of carbon emissions at levels that are too high to reverse global warming even if these targets were met. Yet even these inadequate targets are flouted by many of those who have signed up to Kyoto. The world's biggest polluter, the USA, has refused to sign up at all. Also many of the mechanisms that are being used to implement Kyoto, such as carbon-emissions trading schemes, are unsupportable market-based mechanisms. However it remains an urgent task for all of us to push the question of an effective international treaty on climate change higher up the political agenda.

The alternative to Kyoto presented by the environmental movement is the idea of an alternative international treaty based on 'Contraction and Convergence' (C&C). 'Contraction' refers to the need to reduce CO_2 emissions to a level which would halt global warming and which would become the basis for a budget of carbon emissions. 'Convergence' refers to an allocated quota of emissions to every country based on this budget and the size of its population. Unfortunately C&C shares with Kyoto the principle of carbon trading, or 'pollution for sale', as its method of functioning. It also offers no solution to the sticky problem of the enforcement of such a scheme or its imposition on reluctant governments.

Capitalism cannot solve the problem

The capitalist mode of production cannot resolve the problem of climate change, in fact by its very nature it creates the problem. As socialists, of course, we regard the ending of capitalism and the establishment of socialism as fundamental to tackling the ecological crisis. But we cannot accept, that there is nothing that can be done whilst capitalism remains. Nor that capitalist governments cannot be forced by mass pressure to take measures to protect the environment. Such a conclusion would be to misunderstand the potential mobilising power of the issue. The next 20 or 30 years are going to be crucial for the environmental crisis but there is no guarantee that socialism will be established as a world system in that period. We have to make it central now to our revolutionary programme and perspectives.

Climate change and its consequences will be crucial in the struggle for socialism in the coming period. Wars and conflict over energy resources are built into the process of climate change. There will be wars over oil, over water, over gas. Population movements fleeing desertification or flooding will fall foul of national immigration controls and will face the wrath of the police and the immigration service. The UN estimates that approximately 50 million people are 'environmental refugees', forced to leave their regions of origin in the wake of drought, floods, soil erosion, and the extension of export oriented agriculture. Governments will resort to repressive measures to resist these processes. These are all issues around which the struggle for socialism will be based.

We support the following demands

- Emergency action to reduce the use of fossil fuels alongside massively increased investment in sustainable energy, including solar power, bio-fuel, and wind and wave power; government support for more energy efficient lifestyles. The rich countries must take the lead in this with developing countries taking different, less carbon based, forms of development.
- Tough action against corporate polluters; strict control of all forms of industrial pollution.
- An international treaty that goes well beyond Kyoto in controlling carbon dioxide emissions.
- Global action to help third world countries in sustainable development including debt cancellation and an end to western agricultural subsidies.
- Stop the destruction of the rain forests, defend biodiversity, massive reforestation programmes.
- Curtail activities not essential to human wellbeing, such as the

advertising, sales, arms and many other industries.

- An international plan on the use of water; measures to deal with the increasing problems of drought and floods.
- Public ownership of the oil, gas, coal and water companies.
- Redesign towns and cities to improve the environment and reduce energy consumption: public facilities at manageable distances, end out of town shopping, prioritise walkways and cycle lanes.
- Locally grown food to reduce food miles, and measures to protect the small producers; public ownership of the major supermarkets.
- Cheap and integrated transport systems to provide a real alternative to the car, even in rural areas; publicly owned transport – buses, railways and airlines.
- A halt to road building programmes, particularly in the developed countries, except where improvements would lead to lower fuel consumption and would not lead to more cars; prioritise research into emission free vehicles.
- Halt airport expansion, new runways, and end the fuel tax-break to the airlines.
- Increased public investment to make homes and public buildings energy efficient; new buildings to be carbon neutral.
- High quality facilities to maximise recycling; no new incinerator building.
- No more nuclear power! We oppose nuclear power and all attempts to present it as a clean, safe, carbon neutral, alternative to fossil fuel.
- A halt to the import and export of nuclear waste, and the dumping of it on third world countries.

Environment:
our common future?

François Moreau

The green movements have made an enormous contribution to raising consciousness globally around the ecological questions facing humanity. There is not a single country or region that would come out unscathed if the rest of the world or large regions of it experience ecological collapse. The disappearance of the ozone layer will affect the whole globe, and the same goes for the greenhouse effect.

Indeed, only concerted action on a world scale can allow today's extremely serious ecological problems to be overcome. For example, it is no use suppressing production of toxic chemical elements in one particular country if the multinational chemicals industry just moves production to somewhere where the ecology movement is weaker. The ban has to be universal to be effective.

Even the international bourgeoisie—or, at least, its most enlightened representatives—is beginning to understand this. The last Mohicans of unlimited growth continue to deny the scale of ecological problems, putting their faith blindly in yet-to-be-invented technological solutions, but this obscurantist attitude is now increasingly giving way to a formal recognition of the gravity of the situation. World conferences are mushrooming, whether around the questions of the ozone layer, the greenhouse effect or deforestation. Resolutions are voted, protocol agreements adopted and undertakings made. Western governments are turning green after reading opinion polls.

The image of the Earth as a space ship, in which we are all passengers, has never been so used and so hackneyed. This is not only used as a visual presentation of the situation, but also to push the idea that all classes and nations have a basic common interest in terms of the planet's future. But, far from making social antagonisms obsolete, the threat of a planetary ecological disaster pushes them, on the contrary, to fever pitch. It was similar to the Titanic, where a specific form of class struggle broke out over who was going to get into the lifeboats: 75% of first-class passengers survived, while the same percentage of third-class passengers drowned. It will not be any different in today's situation.

We will start by looking at the first models of the world formulated at the beginning of the 1970s, in order to trace out the evolution of the ecological discussion up until the publication of the Brundtland Report by the UN Environment and Development Commission in 1987, before concluding on current perspectives from an ecological and socialist point of view.

The planetary limits of the Meadows Report

Up to the end of the 1960s, futurists unanimously painted a future of unlimited technological growth, continued economic expansion and a guaranteed increase in living standards for the whole of humanity in the decades and centuries to come. These projections were expressed in the most polished form in a work published in 1967 by Herman Kahn, *l'An 2000* (The Year 2000), which extrapolated the rapid economic growth experienced from 1950-1965 over several future decades.[1]

This smug and apologetic optimism was kicked in the teeth when the Club of Rome published a report in 1972 on growth limits.[2] It was written by a research team from the Massachusetts Institute of Technology (MIT), led by Dennis Meadows. Using simulation techniques borrowed from systems dynamics, the authors' goal was to determine if current developmental tendencies could continue in the future.[3] They arrived at the most categorical of conclusions: unavoidable ecological collapse sometime in the middle of the next century if the exponential growth of population and industrial and agricultural production on a world scale was not halted as soon as possible, and at the latest before the year 2000.

The reasons for the collapse in the Meadows' team projections lie in planetary limits. The Earth contains a determined quantity of nonrenewable resources, it possesses a finite amount of agricultural

75

land and its capacity to absorb toxic waste from industrial processes is limited. Whatever progress is made in the future in terms of recycling resources, agricultural productivity or pollution control, the pursuit of industrial growth will inevitably come up against planetary limits and, still according to the Meadows' team, the whole thing will end in a global ecological collapse. All efforts to prolong the expansion by technological devices can only succeed in putting off the collapse by a few decades at the most, giving it an even more catastrophic character when the inevitable day of reckoning arrives.

Obviously, these conclusions were criticized in the years to follow. A research team at Sussex University in Britain took on the task of going through the MIT group's extremely complex model with a finetooth comb, in order to evaluate its validity.[4] The Sussex team believed they detected a systematically conservative bias in the MIT group's projections at several levels. They also demonstrated that it was enough in most cases to adopt a different, but just as plausible, working hypothesis to that of the Meadows' team to arrive at noticeably different results. This was contrary to the pretensions of the Meadows' team, who wanted their report's global results to remain unaffected by the initial hypotheses in terms of concrete estimations of this or that original magnitude.

The question of nonrenewable resources is one example of this. In its projection, the Meadows' group did not take into account technical progress that could enable previously unusable mineral deposits to be exploited—as has been the case since the beginning of industrialization—and so they implicitly formulated the hypothesis that technical progress was now worn out. According to the Sussex group, the introduction into Meadows' model of a modest, but plausible, rhythm of technical progress (2.3% per year in the domain of extraction technologies), combined with recycling materials, would result in pushing back the exhaustion of natural resources to a point beyond the model's timescale horizon. This is one way in which the final results quite simply reflect the conservative bias of the original hypotheses.

But the most serious critiques made of the Meadows' model do not relate to the technical aspects of the problem. They counterposed, on the one side, humanity en bloc along with its polluting industries, and, on the other, nature in the form of planetary limits. This is particularly clear through the use of world averages and additions, thus making the gulf between the industrial countries at the centre and those of the periphery totally abstract. As for the social relations of production, this is just nonexistent—as if the control of a minority who own the dominant companies has nothing to do with the question of ecological balance.

The incapacity or refusal of the Meadows' team to investigate the social dimension of the question led them to look for a technocratic solution that did not challenge the present relations of domination between classes and on a world scale. This meant that they ended up proposing a solution that is clearly unacceptable for the majority of the world's population, because the halting of economic growth in the next few decades at the latest implies a definitive sanctioning of presentday inequalities. But this "solution" was no more agreeable for the dominant classes in the developed countries, because it would put an end to the accumulation of capital without which the capitalist system could not keep going.

The most important gain of the Meadows' Report is without doubt to have highlighted the need for an overall, longterm analysis on a planetary scale, and so swimming against the tide of the inherent tendencies in modern institutional science to compartmentalize knowledge and split up the objects of research into tinier and tinier segments. On the other hand, the Meadows' team remained imprisoned in a technocratic vision of the world that prevented them from comprehending the social and political contradictions at work on environmental questions.

The second generation of world models

The Club of Rome's second report, published in 1974 by Mihajlo Mesarovic and Eduard Pestel filled some of the holes in the first report and progressed in terms of posing the problematic of worldwide ecological balance.[5] In contrast to the Meadows' model with its global additions, the world model elaborated by Mesarovic and Pestel distinguished ten regions with particular characteristics: four developed capitalist regions; four peripheral capitalist regions; the

USSR and Eastern Europe; and, finally, China. This model tried to grasp the dynamic of development in these ten regions, as well as how they interact commercially, financially and technologically.

The simulation of this model led to a remarkable result: the first region in the world that would experience ecological collapse would be Asia and the SouthEast, including above all India, Pakistan and Bangladesh. After this projection, the widening gap between demographic growth in these countries and their agricultural resources would lead to an increasing food deficit that they would be incapable of making up by imports due to lack of resources, and which would lead to a massive famine around the year 2025. A billion people would die, and eventually a new balance would be restored at much lower population levels.

Apart from the apocalyptic aspect of such a projection, it implies a cornplete overturning of perspectives concerning the relationship between levels of economic development and ecological difficulties. Until then, the general tendency was to think that it would be the most industrialized countries that would experience the worst environmental problems, with pollution, industrial waste, underutilization of water, and so on. Meadows thought that the regionalization of his world model would simply lead to bringing forward the date of the final collapse in the developed countries where pollution is concentrated. And here was Mesarovic and Pestel's world model giving a completely different projection: the first region to suffer a global collapse would be, on the contrary, the poorest in the world alongside black Africa, another region that is the victim of growing ecological imbalance. This should not be surprising. The falseness of the dictum that everyone is equal before death and the tax man—the poor die younger and the rich avoid paying taxes more easily—has been known for a long time. In the same way, any imbalances in the world fall disproportionately on the dominated countries and the poorer classes, while the richest countries are able to devote bigger budgets to correcting the most acute environmental problems, and the dominant classes of all countries have the possibility of fleeing the filth and personally sheltering themselves from toxic fumes on their private islands and their exclusive estates. It is never the possessing classes who suffer from famine. Only the poor die of hunger.

The exporting of ecological problems from the industrialized countries to those in the capitalist periphery received a lot of publicity in 1988 around the odyssey of an Italian cargo ship carrying toxic waste destined for Nigeria. It is common knowledge that, with the growth of environmental protection movements in the industrialized countries, many multinationals have decided to localize the most dangerous and polluting industries in the "third world" where the ecology movement is weaker and where they can count on more repressive governments at the beck and call of the multinationals. Then there was the catastrophe at Bhopal, in India, where a factory belonging to the US chemical firm Union Carbide billowed out highly toxic gas over hundreds of thousands of people living in neighbouring shantytowns. A similar catastrophe happened in Mexico. In general, the media tend to treat ecological disasters as natural phenomena: floods, droughts, landslides.... But these "natural" catastrophes owe a lot to human actions. The recent floods in Bangladesh are linked to the deforestation of Himalayan slopes, the source of the big rivers that cross the country. In its turn, deforestation is a result of pressure from peasants searching for new land and energy sources, wood being the only accessible fuel, particularly since the price of petrol has shot up. Pressure on land is the result of the mechanization of big farms, which chase small peasants away and make agricultural workers redundant. The urban economy and underdeveloped industries cannot absorb them, above all in the context of 1980's austerity policies. The austerity policies and industrial regression imposed for a number of years by the IMF and other agencies of international capital have only made all these problems worse.

Inequality in the face of ecological problems is only one specific manifestation of the relationships of domination that exist on a world scale between the imperialist centre and the capitalist periphery. Mesarovic and Pestel's model empirically seized on this reality and updated its eventual consequences, in the first place in terms of upsetting the balance in the Asian and SouthEast countries. However, they do not draw out the appropriate solutions about the social nature of existing and coming ecological crises. The relations between humanity and nature are not direct but mediated by a mode of production where humanity finds itself in a global hierarchy of dominant and dominated nations and classes. Indeed, in the so-called underdeveloped countries poverty is not an inherent state to be put down to too many inhabitants or to an alleged conservative hostility to technical progress. Bangladesh, for example, has already been richer than England—before becoming its colony, of course. The poverty in these countries is in large part the result of a process of underdevelopment that owes a lot to colonial and neocolonial structures of domination, through which a large part of the resources and riches of these countries were systematically extracted by capitalism in a process of unequal exchange between littleworked resources and manufactured goods. This increases even further the imbalance between the resources of these countries and their populations, particularly as the resources that stay in the country are monopolized by the local ruling classes who collaborate with imperialism. All of this condemns increasingly numerous populations to try to survive on the basis of diminishing resources, with all that this means in terms of deforestation, land exhaustion, overexploitation of mines and so on.

The solution put forward in the Club of Rome's second report was the perspective of balanced "organic growth" between the different regions of the world. According to the authors, this necessitates largescale aid from developed to lessadvanced countries for the latter to be able to cope with their population growth in the decades to come, until urbanization and industrialization lead to a decrease in their birthrates and, in the longterm, to a stabilization of their populations.

At the same time, the Japan group of the Club of Rome, using a world model split into nine regions, proposed that the developed countries gave over around 1% of their annual income to aid without compensation to the "developing" countries, and that this aid be given above all to agriculture and light industry in the peripheral countries so as to increase their agricultural production and provide jobs for their overabundant labour. For their part, the countries of the centre should gradually give up their light industries and put more emphasis on their own agriculture so as to meet their own food needs without importing from the South, as is the case today. All this should allow a balanced development to be followed during the next century and avoid the massive famines projected in the first two Club of Rome reports.

A new international order stillborn

The need for a "new international economic order" has become a major theme since the second half of the 1970s. In 1976, the Club of Rome published a new report prepared by the Dutch economist Jan Tinbergen, which called for better international cooperation and a transfer of financial and technical resources from the North to the South with the aim of achieving the satisfaction of basic human needs. The report differentiated itself a little from the traditional ideology of economic growth at all cost by indicating its habitually destructive consequences in terms of widening social inequalities, and by advocating as a priority the reduction of inequalities between countries and inside each country.

But the main interest of the report is that it went so far as to raise (timidly) the need for big structural changes in order to assure viable development that would satisfy basic human needs. It talked about agrarian reform "where necessary" (where?), and about reducing military spending, considered as a crazy waste of resources in a world where there is so much to do.

The report also underlined the big fluctuations in raw material prices and their general tendency to decrease in relation to those of manufactured products. From an ecological point of view, the low prices of raw materials for industry are nothing to rejoice about. They encourage growing consumption of raw materials and discourage conservation and recycling, thus accelerating the squandering of nonrenewable resources.

A convincing demonstration of this is what happened when oil prices rocketed in 1973-74 and again in 1979-80. Energy conservation suddenly became the number one priority. Although oil consumption has grown more quickly than industrial production since the beginning of the century, it has decreased by a good 25% relative to the volume of total production in the central capitalist countries during the last fifteen years because since then there has been a strong financial incentive to conserve energy. The effects of this were rapidly felt, and we saw a substantial amelioration in energy efficiency in terms of production as well as of consumption.

The best thing that could have happened as much from the

viewpoint of conserving resources as from that of the financial and commercial situation of thirdworld countries—would have been a similar development for other natural resources. Sadly, the union of very specific interests that come together in the case of oil—between the OPEC countries and the US oil multinationals—has no equivalent for other raw materials, which have continued to be sold at bargain basement prices and to be squandered as if there was no tomorrow.

The advocates of unlimited capitalist expansion rejoice about these falling prices, which appear to them to refute any perspectives of a shortage of resources.[7] But the Tinbergen team correctly linked this phenomenon to the functioning of international markets for raw materials, which are buyers' markets for most products. In the capitalist system, the market value of any product comes from the amount of labour necessary for its production or extraction, tempered by the relationship of forces between buyers and sellers and between the working class and the bosses. It is unimportant that nature has taken dozens, hundreds, millions or even billions of years to create these materials, which will be removed in a few hours: that is unimportant from the individual capitalist's point of view, even if it leads to the impoverishment of the whole of humanity through the misappropriation of its natural heritage.

To rectify this situation, the Tinbergen Report foresaw the establishment of a world tax on raw materials, which would be levied by an international body and which would be given over to "third world" countries to finance development projects. Raising the prices of raw materials to industrial users would force them to use them more sparingly and prolong the lifetime of reserves accordingly, in addition to transfering financial resources towards the periphery. It is unnecessary to point out that this project has remained a deadletter because of opposition from the big capitalist powers which import raw materials, the United States heading the list. This has also been the fate of other projects for a "new international economic order", like those formulated in 1980 by the Brandt Commission.

Indeed, towards the end of the 1970s, the political orientation of the Western governments has taken a sharp turn to the right, symbolized by the arrival in power of Thatcher, Reagan, Kohl and, later, Mulroney. Freemarket laws have been worshipped with more devotion than ever. Keynesian (and social democratic) ideas about economic stabilization and income redistribution have been chucked out in favour of a glorification of blind competition and "natural" inequalities. All power to money! The United Nations can carry on chatting about a new international economic order: the real policies are decided at the International Monetary Fund, under the control of the big capitalist powers and for the greater glory of the banks and multinationals.

The estimate made by the Japanese group of the Club of Rome—according to which a transfer of 1% of income from the developed countries to the "third world" would be enough to keep world development on the right track—has already been mentioned. In reality, the transfer of resources happens in reverse, from the periphery to the centre, in the form of repatriated profits, interest and debt repayments which, according to the official statistics compiled by the big international bodies, represent only the tip of the iceberg. It is the direct result of monetarist policies of raising real interest rates that have been carried out since the beginning of the 1980s, and of the radical austerity policies imposed on "third world" countries by the IMF and the World Bank. The capitalist periphery is paying to bail out the centre... exactly the contrary of the Club of Rome's predictions since its second report.

Indeed, the recommendations to transfer resources from the centre towards the periphery are utopian in the context of the world capitalist system, because they go totally against the grain of the system's logical functioning, which has led to a systematic transfer in the opposite direction during the last five centuries.[8] Today, these profits, interests and royalties levied from the periphery represent a significant contribution to the coffers of the dominant multinationals, banks or industrial enterprises, and all the stops will be pulled out to preserve and increase them.

In the same way, the recommendations intended to stabilize the level of development in the most industrialized countries in order to allow the "third world" to catch up are completely abstract in terms of interimperialist rivalries. There is not a single big capitalist power that is ever going to agree to deliberately slow down its own economic growth, for fear of being outdistanced by its rivals. Even if they wanted to, today's supranational bodies such as the UN could never impose such a policy on the big powers, because these latter always have the last word on crucial questions. Only a truly worldwide government could implement a real programme of massively transferring resources from the centre to the periphery, and this would mean first of all tearing state power out of the hands of big capital in the main countries and putting it in the hands of a new government that would rule in the interests of the working masses on a global scale. The complete opposite of the bourgeois United Nations of today.

Ecological crisis and economic crisis

This leads us into asking what future perspectives are available for capitalism. In spite of their utterly different longterm predictions, the analyses of Meadows and the Club of Rome share one common point with those of Kahn and company: the absence of internal blocks in the capitalist mode of production. Thus, production will see unlimited growth, at least until an external limitation puts an end to it. For Meadows, this external limit is represented by planetary resources, and therefore (capitalist) economic expansion will end towards the middle of the next century. For Kahn, this planetary limit does not exist, and expansion will therefore be unlimited, although perhaps following a slower rhythm in the future because of a gradual saturation of basic needs.[9]

The belief in the possibility of unlimited capitalist development—at least within some planetary limits—was an integral part of the intellectual atmosphere of the 1960s. Bourgeois economists were convinced that prolonged periods of crisis, like those in the 1930s, now belonged to the past thanks to the techniques for economic stabilization established in the Keynesian revolution. This was believed to such a point that Meadows, in his world model, treated the years 1900-1970 as a single period of worldwide exponential growth, passing over three decades of virtual stagnation in the better part of the capitalist world between 1914 and 1945. A chance mishap, undoubtedly. Even some authors claiming to be Marxists admitted the possibility of unlimited development, one example being the partisans of the state capitalist theory. Others dealt with the problem simply by denying that any expansion had taken place from 19451970.

However, the Marxist analysis of capitalism's contradictions should have led to the conclusion that it was impossible to support the postwar expansion indefinitely. As early as 1964, Ernest Mandel wrote that: "the long period of accelerated growth is probably going to end during the I 960s".[10] The gradual lowering of the rate of profit from its high postwar levels, the saturation of markets by a growing range of durable consumer goods and the development of excessive production capacity were the main causes for this.

Within the regime of private ownership of the means of production, profit is the only motive for producing and investing, and this generates and necessitates everincreasing capital accumulation. This is why the capitalist system can never remain stationary: it must grow or perish. Nevertheless, the system is incapable of sustaining longterm growth; it always ends up by getting bogged down in its own contradictions, as happened between the two world wars and at the beginning of the 1970s. Capitalism has to grow in an unlimited way, but it is incapable of doing it. The result is periodic crises that are increasingly profound and destructive, dangerous not only for humanity but also, increasingly, for the ecosystems of our planet.

Today, there is a consensus among analysts that the capitalist system entered a period of slower growth from the end of the 1960s (or the middle of the 1970s, according to different authors), although the causes of this phenomenon are more controversial. There has been a rediscovery of theories on longterm economic movements and "Kondratieff waves", which had been relegated to economic museums for over thirty years. But what are the consequences of this from the point of view of longterm ecological balance?

From a "naive" ecological point of view, if one can use the term, the marked slowing down of capitalist expansion should be a cause for celebration. Is this not exactly what the first Club of Rome report extolled? Except that this slowing down is the result of socioeconomic mechanisms completely absent from the Meadows' Report, and in no way from the application of its recommendations. Effectively, the entry into a period of slower growth has already meant that the consumption of raw materials is much less than foreseen in Meadows' Report. Indeed, several metals that should have already been exhausted according to his predictions are still abundantly available, tin for example. But going from that to saying that the slowing down of capitalist growth is beneficial for the environment is a step that should not be taken.

The slowing down of growth is not the result of a conscious and well thought out decision by anybody, but of a structural crisis whose effects weigh on the dominated classes and peoples. Increasingly big layers of pauperized masses in the capitalist periphery are reduced to living on or below survival levels. This increases the burden on the environment, notably in terms of pressure on land and forests. The forced opening to the World Bank market leads many countries onto the path of industrialization that is unplanned arid destructive, both for human beings and for the environment. The draconian budgetary restrictions imposed by these same policies compromises even further the meagre resources consecrated to defending and protecting the environment.

In the developed countries, the crisis has caused ruling groups to slide to the right in their search to reestablish their rates of profit by stepping up attacks on workers' living standards. Some classblind ecologists could celebrate the consequent restrictions of mass consumerism, and even actively support policies in this direction. They forget that nearly half of goods and resources are consumed by the richest 20%, mainly the bourgeoisie and the pettybourgeoisie, while the poorest 50% have to share 20% of the cake, which gives them each truly minuscule portions. It is not here that waste is to be found, but it is here that austerity policies hit the hardest, leading to an even more unequal distribution of incomes, and therefore to an even bigger waste in luxury consumerism of the ruling classes. The unbelievable wealth of deluxe car manufacturers in the 1980s is a striking example.

More generally, the restoration of complete primacy for private enterprises and unbridled competition goes directly in the face of efforts to regulate in terms of the environment, from pollution to toxic waste passing by the overutilization of resources, under the pressure of competition. Without mentioning the relaunching of the arms race, which is an integral part of the rightwing's programme and leads to monstrous diversion of precious financial, human, technical and material resources towards utterly destructive ends.

Victims are guilty of overpopulation

The slide to the right of bourgeois policies during the 1980s has also been reflected ideologically by the rehabilitation of classical Malthusian theories, according to which uncontrolled world population growth constitutes the ultimate cause of the risk of ecological catastrophe. This is accompanied by a barely concealed modem version of the "yellow peril", because this growth is concentrated today in the countries of the capitalist periphery, as against the virtual demographic stagnation in the imperialist and industrial countries generally. The Malthusian theory corresponds exactly to the relations of domination played out globally between the dominant bourgeois classes in an imperialist centre comprising the developed capitalist countries on the one hand, and the poor masses in the capitalist periphery made up of the dependent countries in the socalled third world on the other.

International capitalist bodies like the World Bank generally attribute the poverty of these countries to their overpopulation, from which follows in turn the lack of education, the traditionalism and so on. This contemptuous vision of the dominated peoples ignores the real reasons for high birth rates. One cause, paradoxically, is the high rate of infant mortality (more children have to be born to ensure that some will reach adulthood). Others include the absence of social security and oldage pensions, causing older people to be dependent on their children; the maintenance of women's oppression, which prevents, them from controlling their own bodies and reproductive functions; and economic insecurity, which forces families to depend on a number of incomes and, in particular, the work done by children. In these conditions, there is nothing surprising in the fact that campaigns to plan birth rates end in resounding failure: people want children because they need them!

But once social structures are radically changed, all this will change very rapidly. The most striking example is in Cuba, where the revolution led to a noticeable improvement in living conditions in terms of nutrition and healthcare, linked to a big increase in both education and employment. The revolution also appreciably improved the status of Cuban women. Older people could now count on pensions. The most remarkable thing was that all this caused demographic tendencies in Cuba to come into line with those in the most advanced capitalist countries in just a few decades, to the point where now a stabilization of the population at a slightly lower level is predicted. This improves both the possibilities for increased living standards and the quality of life in the island.

China is another example of a radical change accomplished through revolution. In 1948, a Malthusian writer named Williams Vogt had predicted the inevitable extermination of tens of millions of Chinese people (out of a population at the time of 400 million), either by famine, war or epidemic.[11] The 1949 revolution took place and China was now feeding 1,100 million inhabitants, in conditions of health better than nearly all the other third world countries and similar to those in many developed capitalist countries. That was made possible by the elimination of the landowning class, which monopolized an enormous proportion of the land, with land being redistributed to the peasantry. The improvement was very marked, outside of a number of political errors committed by the Chinese leadership and without which progress would undoubtedly have been even bigger.

One of these errors had precisely consisted in taking completely the opposite course to the Malthusian theory, according to which all the problems stemmed from socalled overpopulation. The Chinese leaders concluded that the population was not and never would be a problem; to the point of launching policies aimed at increasing the birth rate in China as quickly as possible. The maxim that each supplementary mouth also brings two additional arms was often quoted. Unhappily, they did not also bring more cultivable land. The belated realization of this truth led to a brutal aboutturn in demographic policies in favour of an authoritarian limit on births, the famous "one family, one child" policy. This could have been avoided if, from the point of view of ecological balance, the existence of an optimal population level in relation to a given technological level had been recognized, in place of blindly following a policy of maximum population growth as if things would always turn out alright. This lesson must not be lost for future revolutions.

Ecological crisis and social relations

The global scope of ecological problems and their social dimension were expressed most solemnly in the Brundtland Report, produced by the UN commission on the environment and development.[12] The importance of this report lies not only in the fact that it represents an official recognition of the urgent need for concrete action on a worldscale to avoid the numerous dangers facing the environment, but also in its recognition that these dangers are rooted in existing socioeconomic structures. However, all this is shrouded in the muffled language typical of international bodies and barely goes beyond the level of a liberal or socialdemocratic criticism of inequalities of income and power.

All the same, this was enough to provoke negative reactions from official employers' bodies. Thus, the financial editor of the *Globe and Mail*, the main bosses' newspaper in Canada, recently assailed the Brundtland Report for its supposedly socialist recommendations, impregnated with United Nations egalitarianism and aiming to introduce the kind of centralized planning that had bankrupted the East European countries![13]

But there is no reason to worry about their point of view, because the supreme political authority still remains at the level of nation states, and the international bodies do not have any real power in the face of the strongest states. The only governments that international bodies are in a position to compel to do anything are precisely those in the weakest countries, which are in a position of commercial, technical and financial dependence in relation to the big capitalist countries. And this power is used unrelentingly to force them to adopt anti-worker, anti-popular and anti-women austerity policies that only serve to aggravate the ecological imbalances in these countries.

The governments that should most be compelled to follow responsible ecological policies are precisely those over which international bodies have the least authority—those of the big capitalist countries. Indeed, the international bodies are dependent on them. The United States only has to suspend paying their dues to the UN to block decisions that they do not like.

The basic problem here resides in the fact that state power in the big capitalist countries represents the interests of big capital in each of the countries concerned in terms of its competition with the big capital of other countries. But intercapitalist competition is not decreasing, but increasing under the effects of the structural crisis of the world capitalist system since the end of the 1970s and the growth of European and Japanese capitalism that is threatening the dominant position of the American bourgeoisie in the world. In these conditions, one might as well believe in Father Christmas as hope to see wide-scale concerted action come out of international bodies dominated by these rival powers.

The ecology question dramatically confirms one of the basic principles that the Fourth International has always defended: the global character of the class struggle and the need to build a worldwide instrument to properly lead this struggle, against capitalism as well as against the bureaucratic regimes in Eastern Europe. This is what motivates the Fourth International's efforts to build a world party of socialist revolution that unites all forces active in every country inside the same movement. Only such a world perspective opens the possibility of a fundamental and lasting solution to ecological problems.

"Think globally, act locally" is one of the most well-known sayings of the ecology movement. But for thinking globally there is nothing better than a world movement that integrates contributions from its local groups everywhere, in all countries, in all regions. As for local action, to be useful it has to be part of an overall strategy that is conceived and implemented on the same scale as the problem is posed—that is, on a world scale.

All together?

In fact, the Brundtland Report—as its title clearly shows—is a plea for class collaboration on a world scale and between the social forces in each country, on the basis of the supposedly common interest of all nations and all social classes in preserving the environment.

This appeal for class collaboration and international cooperation hardly pleases the imperialist hawks and the conservative Western dinosaurs, who aspire rather to unilaterally impose policies that conform to their interests by exploiting their position of strength in the system. But the green thematic can easily be taken over by classic bourgeois politicians, who read the opinion polls and know that environmental problems have become one of the main preoccupations of the masses in the Western countries, as proved by the success of the European Green parties.

The most intelligent bourgeois forces have also understood for a long time which parts of the ecology question they can use as green justifications for their capitalist austerity policies and to undermine workers' demands. What, you want to increase your consumption when the world's resources are already running out and when the majority of the planet's population already lives in abject poverty? As always, class collaboration requires the working class dropping its own demands, because it is not the bourgeois class that is going to give up defending its own interests!

However, the search for class collaboration to save the environment comes up against the existence of fundamentally divergent class interests on this question as on others. The ultimate root of ecological degradation is to be found in the search for private profit, based on the capitalist mode of production. This creates a contradiction between the short-term profit of private firms on the one hand, and the long-term interests of humanity on the other. This contradiction becomes clear if we look at things from the point of view of a company that uses toxic chemical products. It faces two choices: either to use these products in such a way that they do not harm the environment, which puts up costs, or to throw them in the river, which costs little. There is no point in spelling out which option will be chosen. And it is not even a question of bad will: it is one of economic constraints for private businesses, because if they do not throw their toxic products in the river, their competitor will.

Here, we find ourselves in the presence of what neoclassical economists call "externalities"—that is, the costs or benefits that a given firm brings to society without society having any say in it. It is one of the rare cases where the adepts of traditional free-market economic policies accept the principle of governmental control, when the external costs for society exceed companies' internal benefits, pollutant in this case. This is at least the case for the classical free-marketeers, because the "libertarians" of the modern right prefer a "market" solution, with the victims of pollution paying the polluters not to pollute... logical, isn't it?

But control within a capitalist regime is never a real solution. Of course, we are for strict control over polluting industries (to take one example): we are in favour of a rigorous application of these controls. But, at the same time, when the financial incentives for polluting are so overwhelming, their ultimate ineffectiveness has to be recognized. Unless a government inspector is allocated to every industrial entrepreneur using toxic products, with a second inspector behind the first to make sure he does not start taking bribes. It is easy to see what sort of bureaucratic monster such a system would lead to. In any case, we are a long way from something like that. In Quebec, the environment minister's "Green Police" have around 60 inspectors to supervise the use of around 30,000 toxic products in 3,000 different workplaces.

Moreover, the capitalist governments themselves and their state companies are often among the worst enemies of the environment, because in the last analysis they are also subject to the same logic of private profit that drives the whole of bourgeois society. For example, it is the state-owned national company Hydro-Quebec that is the real culprit of the catastrophes of BPC at Saint-Basile le Grand, without mentioning James Bay or the Gentilly nuclear power plant. At most, governments have limited the damage via subsidies or patching things up here and there when public indignation has become too strong and too electorally dangerous.

But there are inspectors already installed in every workplace that is a potential or actual polluter in every country: those who work in them. They know what is going on; they know what risks are taken with the environment, because they are the daily agents of capitalist production. But they are agents subordinated to the control of those who own the means of production, whether they represent private interest groups or the capitalist state.

To tackle the problem at root it is necessary to wrest control of the workplaces from private capitalist interests and hand it over to the workers in each enterprise, in the framework of democratic management of the whole of the economy by the whole of society. This is the only way to ensure that the production methods used correspond to the real and long-term interests of the working majority, and not to the private short-term interests of a capitalist minority which carries out a policy of *après moi le deluge,* just like the aristocracy before the rise of capitalism. And it will literally be the deluge, with the greenhouse effect, which will cause the ice floes in the Arctic and Antarctic to melt, with the oceans rising by several metres and all the coastal regions of the planet being submerged.

So the ecology question provides undoubtedly the most striking contemporary illustration of the historical alternative formulated at the beginning of the century by the German revolutionary Rosa Luxemburg: socialism or barbarism. Either the working class will succeed in establishing its democratic control over the economy and society, restructuring them in relation to human needs in a long-term perspective, or capitalism will wreck the future of humanity on this planet. And the days of reckoning are getting ever closer!

The counter-example of the Eastern European countries

Undoubtedly, some people will retort: does not the present situation in the USSR, China, Poland and the other so-called workers' states prove that ecological problems are not caused by capitalism as such, but simply by industrial society? Indeed, ecologist currents rejecting both capitalism and socialism as two equally harmful forms of production can easily use the example of the USSR, where some of the worst ecological disasters of contemporary history have occurred. But the USSR today and other similar regimes are precisely not models of the socialist society that we are fighting for. They are the product of a proletarian revolution that overthrew the power and the property of the bourgeois class in these countries, but where the working class did not succeed in installing or preserving its own democratic control over society. Real control fell into the hands of layers of privileged bureaucrats who hold onto power by every possible means, including using violence against the mass of workers and people if necessary, assuming a total monopoly at all levels—economic, political, ideological and at the level of information. These are ideal conditions for developing blind productivism and wild industrial gigantism, utterly destructive for the environment and totally divorced from the satisfaction of real human needs. This is exacerbated by the fact that the counterweight that can exist in the capitalist countries for putting on the brakes or stopping this or that anti-ecological project or imposing this or that progressive measure does not apply (or at least has many more difficulties) in these societies, where the bureaucracy represses all types of movements.

But the Fourth International was founded precisely against this type of bureaucratic regime, in the struggle to overthrow them and to restore socialist workers' democracy in these countries, at the same time as it fights for overthrowing bourgeois power in the capitalist countries. This struggle for socialist democracy that many people think is utopian is now a burning actuality in countries like China, Poland and the USSR, and the perspective of overturning the ruling bureaucracies is now a concrete possibility.

It is interesting that the growth of movements for democratization in these countries is often accompanied, if not nourished, by increased consciousness of the ecological dramas that are unfolding and the need to stop them, given the total irresponsibility of the ruling bureaucracy. Many victories have already been notched up, such as the decision by the Communist Party of the Soviet Union's political bureau to cancel an insane project to divert rivers in Siberia, or the Hungarian government's decision to cancel a dam project on the Danube following mobilizations by the green movement.

This relates to another fundamental aspect of the Fourth International's programme as against other currents calling themselves Marxist, which is its insistence on the need for socialist workers' democracy. Contrary to the point of view extolled by Stalinist currents, this is not some pettybourgeois luxury that we can do without, if not a nuisance for "building socialism" by an enlightened leadership with the right line. On the contrary, it is vital for preserving the future, ecologically as in other spheres, by assuring that political power is really in the hands of the working masses in terms of its actions and priorities, instead of being imposed on them for better or worse, too often for the worse.

The Marxist critique and the socialist alternative

Capitalism's incapacity to assure harmonious and sustained development of the productive forces has always been one of the main arguments raised by Marxists to justify its overthrow, and developments over the last 15 years have once again illustrated this. But going from that to preaching the most rapid possible development of production as the principal objective for socialism, if not even its content, is a step taken only too lightly by most of the forces calling themselves Marxist.

Since Marx' and Engels' polemic against Malthus, in general Marxists have firmly placed themselves in the camp of technological optimists, convinced of the unlimited possibilities of science. Marxism proudly claimed for itself the struggle begun by the industrial bourgeoisie for the submission of nature to "Man" with a capital M, in an unreserved scientism. One of the main critiques directed at capitalism resided precisely in its presumed incapacity to develop the production forces beyond a certain level. The socialist revolution was therefore necessary... to open up the path of unlimited development of production under socialism.

This distortion of the original socialist project became all the more ingrained as the first proletarian revolutions happened in relatively backward countries and not in the most advanced ones, contrary to Marx and Engels' expectations. The new workers' states therefore found themselves face to face with much more powerful imperialist states leading them to develop their productive base as quickly as possible to the detriment of other preoccupations, which in the circumstances appeared secondary. Stalinism did the rest in terms of defiguring socialism's initial objectives and replacing them with the cult of production targets. Of course, Marx, Engels and other great Marxist writers certainly wrote material that distanced them markedly from the total productivism and scientism described above. But there is no denying that this really did represent the dominant vision among Marxism's best representatives in the twentieth century, including, of course, the Fourth International. This is what ecology movements have come to challenge by raising consciousness about the natural limits of the planet's ecosystem and by compelling the socialist project to be reformulated in the light of the undeniable ecological constraints imposed on us.

But the need to revise the scientist and productivist version of socialism—that has been propagated during the twentieth century provides the perfect occasion to revive the initial socialist project, that of Marx and Engels. Because the idea was precisely not that of promoting production for production's sake, or putting an equals sign between happiness and material consumption. The idea was to allow human beings to fully develop their potential as people by emancipating them from the dominant relations of production and the oppression belonging to class society, themselves based on the material constraints caused by shortages.

Marx was indeed convinced that technological progress was going to allow the full satisfaction of human leads, liberating humanity from the monotonous, repetitive, alienating and exhausting work characteristic of capitalism. After going beyond a certain point corresponding to the satisfaction of human needs, further growth of productive forces would no longer result in expanding the volume of production but in the reduction of working hours and an increase in leisure time, until ultimately production would become completely automated. Humanity would then emerge from under the rule of necessity into one of liberty.

In the face of incarnations of "actually existing socialism", we have to go back to the initial socialist project, which wanted to emancipate humanity from material constraints and abolish shortage, thus laying the material basis for a classless, free and totally egalitarian society. Marxism is only productivist in the sense that it seeks the maximum development of human labour's productive force, precisely with the goal of liberating humanity from work itself. This does not necessarily mean the maximum development of the physical volume of material production: it means satisfying basic human needs that are biologically and socially determined. Beyond this point, there is barely any reason to increase material production again, and subsequent technological progress can be consecrated to reducing working hours even further to increase leisure time, through to the total automation of work.

The recent breakthroughs in microelectronics and robotics have given these projections new credibility on the technological level, but their introduction under capitalist control in a market economy framework completely reverses their liberating potential to turn them into additional instruments of subservience and alienation, as well as causes of unemployment. But it is no good reacting to these

96

developments by relapsing into Luddism, a movement of English artisans who destroyed machinery at the beginning of the nineteenth century. It is necessary to achieve is social control over technology thanks to workers' power in such a way as to draw out all its liberating potential.

But it is necessary to complete this classic Marxist perspective by rethinking the models of consumption and living conditions constructed by advanced capitalism. These are centred notably around individual transport in the shape of the car, the house in the suburbs or the skyscrapers in the city centres, which are all among the most antiecological modes of life that could be imagined. They are impossible to generalize to the whole planet, and even impossible to maintain in any lasting way in the developed countries alone. A massive restructuring will have to be undertaken: developing social housing, spreading out commercial or industrial activities, replacing individual ears by public transport, bicycles and—why not—walking!

Ecological imperatives also mean the pure and simple banishment of a large number of products, techniques, production methods and even whole industries that are intrinsically dangerous and unviable in the longterm, the first on the list being the nuclear industry. From this flows the necessity for an energy strategy based on renewable sources and no longer founded on using up nonrenewable resources such as oil. Even the whole of modem agriculture has to be reviewed, because it rests on the massive use of nonrenewable resources in the form of fertilizers and pesticides, which exhaust soils over time. A considerable amount of research is needed in order to develop viable alternatives that are both ecologically sound and valid from a socialist point of view.

It is true that there are often tensions between the tradeunion movement's immediate demands and those of the ecology movement, above all when the latter presses for the closure of a polluting factory or for banning a dangerous product, whose production means jobs for thousands of workers. These contradictions can only be resolved in the long term on the basis of complete economic reorganization, which plans for the reconversion of dangerous factories and products towards activities that are ecologically viable. But it is hardly surprising that workers want to hold onto their jobs, even if they are dangerous, because even that is better than unemployment!

Everybody has heard of the fashionable theses according to which class struggle is no more and the workers' movement has become anachronistic, if not retrograde, relative to new social movements (including of course the ecology movements) which concentrate on those things that are really at stake in the "postmodem" era. This type of thinking is quite widespread in the green movement itself, which has put up with incomprehension, if not hostility, from the big bureaucratized workers' parties and unions to ecological demands while fighting for the legitimate interests of workers working in polluting or dangerous industries. Did we not hear the leader of the Canadian green party categorically reject any support for the socialdemocratic NPD during the last federal elections in constituencies where there was no green candidate, because of the NPD's links with the unions, who were surely going to oppose the closure of polluting factories?! But we should not lose sight of the face that the global restructuring of productive and consumer relations that is necessary for the survival of humanity can only be established by a political power decided in this manner. And there are not any number of social forces in a position to overthrow the bourgeoisie in contemporary society. There is only one: the working class, men and women in alliance with the other exploited and oppressed layers in society. Far from challenging this central thesis of Marxism, the ecological crisis gives it a new relevance today. Certainly, it is also necessary that socialism becomes ecologist, and the Fourth International has put this point on the agenda of its next World Congress with the aim of updating its programme on this question, profiting from the important gains made by the ecology movement. But it is also necessary that the ecology movement becomes socialist, because this remains the only hope for humanity.

In the short term, in the years to come, in the centuries to come, we have to fight to preserve humanity's future on this planet in the face of the ecological devastations of capitalism in crisis. But we also have to reconsider the classical socialist project so that it also becomes an ecological project, for a world that aims to satisfy human needs by respecting the world's basic natural balance. Classical socialist demands in favour of improving living and working conditions for the toiling and popular majority have to be rethought in terms of models of consumption, looking for new ways of life that put the stress on quality rather than quantity. The technological resources wasted by military production have to be used to develop and refine products, techniques and procedures that are safer, less energyconsuming and more economic in their use of resources. In brief, the transition to socialism has to be reformulated in the sense of restoring a viable longterm balance between humanity and nature.

NOTES

1   Herman Kahn and Anthony Wiener, *l'An 2000*, Marabout, Brussels, 1967. Kahn, a big pioneer of computer nuclear war games, was supposed to have been inspired by the character of Dr Strangelove in Stanley Kubrick's famous film of the same name.

2   Dennis Meadows, Donella Meadows, Jorgen Randers and William Behrens, *The Limits to Growth*, New American Library, New York, 1972.

3   Systems dynamics seeks to simulate complex processes by quantifying the retrospective effects and interactions between the variables involved in these processes, in such a way as to project their most likely future evolution. These calculations can easily include hundreds of equations and thousands of parameters, and are done on computer, which enables the impact of different changes on the evolution of the phenomena being studied to be assessed.

4   Hugh Cole, Cristopher Freeman, Marie Jahoda, Karel Pawitt, *l'Anti-Malthus*, Seuil, Paris, 1974.

5   Mihajlo Mesarovic, Eduard Pestel, *Stratégie pour Demain*, Scull, Paris, 1974.

6   Jan Tinbergen (ed.), *Reshaping the International Order*, Dutton, New York, 1976.

7   A particularly strident exposé of this blind technological optimism and apologetic for capitalism can be found in the last work of Herman Kahn (in collaboration with Julian Simon), *The Resourceful Earth*, Basil Blackwell, Oxford, 1984. Here it is seriously argued that nuclear energy is among the safest of human activities, and that the quality of water in the Great Lakes

is now "excellent" for fishing and swimming.

8   On this, we owe a lot to the work of Immanuel Wallerstein: see *Capitalisme et Economie-monde*, Flammarion, Paris, 1980, in two volumes.

9   Herman Kahn, William Brown and Leon Martel, *The Next 2000 Years*, William Morrow, New York, 1976. This is an attempt to refute the Meadows' Report which wallows in the most delirious technological optimism. The authors are all linked to the American military apparatus, other than Kahn who is known for his contribution to nuclear war games. Brown is a specialist in military strategy, and Martel is described as someone who has worked in military and political intelligence.

10 Ernest Mandel, *l'Apogée du Néocapitalisme et ses Lendemains*, in Traité d'économie marxiste, Vol. 3., 1018, Paris, 1969, p.282.

11 Williams Vogt, *Road to Survival*, New York, 1948.

12 *Our Common Future*, Oxford University Press, 1987. Gro Harlem Brundtland was the social-democratic prime minister of Norway until October 1989.

13 Terence Corcoran, *Brundtland Message Lacks Economic Base*, The Globe and Mail, Toronto, June 14, 1989.

Fourth International
resolution on climate change

GIVEN:

- That it is well established that global warming is in its majority the result of emission of greenhouse gases, derived mainly from burning of fossil fuels as well as land management (deforestation, intensive agriculture, poor soil management, etc.);

- That according to the IPCC, a reduction of at least 60% in greenhouse gas emissions is necessary between now and 2050 in order to prevent major climatic dislocation with incalculable consequences;

- That the most recent available data on atmospheric concentrations of CO_2 and CO_2 equivalents show that we have already entered the lower part of the dangerous fork (450-550 ppmv of CO_2 equivalents), with accelerating rises of atmospheric concentrations of the gases involved;

- That climate change is already making its effects felt, especially on workers and the disinherited masses, in particular in the dominated countries;

- That in the 50-100 years to come these changes threaten to subject hundreds of millions of human beings to the perils from the rising sea levels, the spread of certain diseases, falling agricultural productivity in many regions, declining biodiversity and shortage of water resources (leading to up to three billion victims in 2100 without voluntarist climate policies);

- That faced with these challenges, the capitalist management of climatic disasters and threats (in particular, Katrina in New Orleans, and the threat of rising ocean levels to Pacific islands and other regions) gives cause to fear that imperialism will resort to Malthusian and militarist policies characterized by barbarism on a unprecedented scale;

- That the Kyoto Protocol objectives are totally insufficient for dealing with the danger, and that its objectives have been reduced still further by its rejection by the US as well as the mechanisms of flexibility, which risk having more and more negative by-products, both on people's right to development (the 'low hanging fruit' effect) and on biodiversity (carbon sinks);

- That the economic competition and strategic rivalry among imperialist blocs risks leading to an even worse compromise than the Kyoto Protocol in terms of the fight to save the climate ('voluntary commitments', no commitments, no deadlines), people's right to development, or ecology in general (nuclear energy);
- That, due to the US and Australian refusal to ratify it, Kyoto, even if carried out in full by its signatories, would bring a 1.7% emission reduction for the developed countries as a whole (EEA report, N°8/2005, page 9);
- That the technical potential of renewable energies (direct or indirect solar and geothermal) is the equivalent of 6 or 7 times the current world energy consumption and makes it perfectly possible to avoid major climatic disasters while satisfying human needs and preserving the environment;
- That we reject nuclear power as an alternative: it is expensive and highly dangerous – and it is not carbon neutral;
- That climate stabilization (a 2° maximum increase in T° compared to the pre-industrial era) requires a vast energy revolution combining, in particular, 1) a transition to renewable energy independently of surplus costs, 2) massive reduction of primary energy demand in developed countries, and 3) massive transfers of 'climate friendly' technology to developing countries;
- That this issue as a whole confronts the workers' movement in general and revolutionary Marxists in particular with a series of new tasks and major programmatic and strategic challenges;

THE IC[76] DECIDES

- To take part in unitary mobilizations to save the climate, particularly those that are developing following the appeal from the London Social Forum. In particular we mobilise for the worldwide demonstration on climate change called for from the Caracas WSF, which will take place in November 2006. To this end we participate in the organising committee for this demonstration in Frankfurt on March 4 2006 at the ESF organising meeting;

- To devote more attention to the climate issue and the politics of climate, notably in the press of the sections and the international;
- To devote the 'ecology seminar' decided on by the World Congress to analysing climate change and its implications, in order to elaborate a programmatic orientation and political line on these matters. To this end, the IC calls for the formation of an international network of comrades with knowledge of the various scientific disciplines involved, so as to produce one or more working documents on the theme 'Energy Revolution and Social Transformation'; and
- To put the question on the agenda of its meeting in one year's time.

Part three: popular ecology

The ecological dimension in the 'South'

Fourth International[77]

It would be very mistaken to think ecological issues only concern the countries of the North – a luxury for wealthy societies. More and more, social movements with an ecological dimension are emerging on the periphery of capitalism, the 'South'.

These movements are reacting to deepening ecological problems in Asia, Africa and Latin America, a consequence of imperialist countries' deliberate policy of 'exporting pollution', and the unbridled productivity demanded by 'competitiveness'. We are witnessing the appearance of popular mobilisations in the South in defence of peasant agriculture, communal access to natural resources, threatened with destruction by the aggressive expansion of the market (or the State). Other struggles are arising to fight the damage to the immediate environment brought about by unequal exchange, dependent industrialisation and the development of capitalism (agribusiness) in the countryside. Often, these movements do not define themselves as ecological, but their struggle still has an essential ecological dimension.

It goes without saying that these movements are not opposed to improvements made by technological progress. On the contrary, the demand for electricity, running water, proper sewage and more medical dispensaries ranks high in their list of demands. What they are refusing is the pollution and destruction of their natural surroundings in the name of 'market laws' and the imperatives of capitalist 'expansion'.

A 1991 text by Peruvian peasant leader Hugo Blanco (of the

Fourth International) is a remarkable expression of the meaning of this 'ecology of the poor'. "At first glance, defenders of the environment or conservationists seem like nice, rather eccentric fellows, whose main goal in life is preventing the extinction of blue whales or pandas. The common people have more pressing concerns, for example where their next meal will come from. (...) However, in Peru there are a great number of people defending the environment. Of course, if you told them 'you are ecologists', they would probably answer, 'ecologists, my eye' (...) And yet: who can deny the inhabitants of the town of Ilo and surrounding villages, struggling against pollution caused by the Southern Peru Copper corporation, are defending the environment? And isn't the Amazonian population totally ecologist, ready to die to defend their forests from pillage? Or the poor population of Lima, protesting tainted water?"

Brazil is among the countries where the link between social and environmental issues has been made on a mass scale. We can see the Landless Peasants Movement (MST) mobilising against GMOs, in a direct confrontation with the major multinational Monsanto. Municipalities and provinces governed by the Workers Party (PT) are attempting to make ecological aims a part of their participatory democracy programme. The Rio Grande do Sul provincial government, close to the MST (and the PT), wants to ban GMOs from the region. Wealthy landowners in the region are indignant, going on record against what call an 'archaic outlook'. They view the struggle against transgenic seed as a 'conspiracy to impose agricultural reform'.

Indigenous peoples, living in direct contact with the forest, are among the primary victims of the 'modernisation' imposed by agrarian capitalism. As a result, they are mobilising in many Latin American countries to defend their traditional way of life, in harmony with the environment, against the bulldozers of capitalist 'civilisation'. Among the countless manifestations of the Brazilian' ecology of the poor', one movement has stood out as particularly exemplary, by its social and ecological, local and planetary, 'red' and 'green' scope. Namely, the fight of Chico Mendes and the Coalition of Forest Peoples in defence of the Brazilian Amazon region, against the destructive appetites of major landowners and multinational agribusiness.

Let us briefly recall the major events in this confrontation. Chico

Mendes was a trade-union activist, with ties to the (CUT) and the Brazilian Workers' Party (PT). Explicitly referring to socialism and ecology, in the early 80s, Mendes organised land occupations by the seringueiros, peasants who lived by tapping rubber trees, against latifundistas who were sending in bulldozers to cut down the forest and replace it with grazing lands. Afterwards, he succeeded in bringing together peasants, farm workers, seringueiros, trade unionists and indigenous tribes – with the support of rank-and-file Church communities – in the Alliance of Forest Peoples, that was able to thwart many clear-cutting attempts. International awareness of these actions warranted him the Global Ecological Prize in 1987. However, a short time afterwards, in December 1988, latifundistas exacted a heavy price for this ecological struggle by having hired killers murder him.

Given the links forged between social and ecological struggles, peasant and indigenous resistance, survival of local populations and safeguard of a global imperative (protection of the last major tropical forest), this movement can become a paradigm for future popular mobilisations in the 'South'.

Cuba's green revolution

Dick Nichols

What can Australian environmentalists learn from Cuba, a country that still flirts with nuclear power, is besieged by many environmental problems typical of the Third World, and lags behind countries like Denmark and Holland on issues like recycling, green taxes, alternative energy and eco-labelling?

During a recent visit to "the fairest island ever revealed to human eyes" (as Christopher Columbus described Cuba), I searched for the answer. I wanted to understand the impact of the "Special Period in Time of Peace" – the emergency program to save the socialist revolution after the collapse of the Soviet bloc.

After talking to environmental scientists, administrators and activists, and reading recent Cuban writings on ecology, it is clear that there is a lot of debate about how to reverse environmental degradation. It is also obvious that few Third World countries can match the legislative, planning and educational efforts that Cuba is applying in its battle for environmental sustainability.

Moreover, few environmental movements can match Cuba's revolutionaries in government, scientific institutions, education system and emerging non-government organisations in their passion and dedication to the environmental cause.

For centuries, Cuba's natural resources and beauty were sacrificed to Spanish colonial landowners and, later, US corporations. In the early 1800s, the great Prussian geographer Alexander von Humboldt was already lamenting the destruction of Cuba's native forests.

In his book *Dialectics of Nature*, Frederick Engels – Karl Marx's collaborator – could find no better example of the impact of capitalist greed on the ecosphere than the operations of Cuba's Spanish planters "who burned down forests on the slopes of the mountains and obtained from the ashes sufficient fertiliser for one generation of highly profitable coffee trees ... what cared they that the heavy tropical rainfall afterwards washed away the unprotected upper stratum of soil, leaving behind only bare rock!"

Through such vandalism, Cuba was transformed into an exporter of sugar, tobacco and coffee. Total forest cover fell from 85% in 1812, 54% in 1900, to 14% by the time of the 1959 revolution. To this crime against nature before the revolution can be added many others, including: rapacious nickel mining (coating a wide expanse of the island in red dust); endemic problems created by monoculture crops; and the gamut of damage that goes with rural poverty.

After the revolution

The revolution and the later development of Cuba's economy as part of the former Soviet bloc was double-edged. The revolution eliminated poverty, unemployment, landlessness and illiteracy and built up basic rural infrastructure, thus attacking the degradation of the countryside at the source. Through sweeping land reform, the leaders of the revolution disproved the myth that degradation is due to the pseudo-explanation, still favoured by World Bank functionaries, of "rural overpopulation". For the first time, and despite continuing population growth, deforestation in Cuba began to be reversed. By 1997, the island's forested area stood at 21.5%, a 7.5% increase since 1959.

On the other hand, the model of industrialisation that Cuba adopted in the 1970s generated (when combined with the continuing reliance on sugar exports) a new set of environmental stresses. Oil spills, coastal erosion, rising salinity, algal blooms and high levels of industrial pollution showed Cuba was paying a high environmental price for industrialisation.

Even though environmental protection featured strongly in the country's law books, the impact on factory managers was often minimal. According to Cuban environment teacher and writer Carlos Jesús Delgado Díaz: "A study carried out by the National Assembly of People's Power at the end of the 1980s reflected the fact that, when faced with the choice of fulfilling the production plan or breaking the law, a significant number of administrators plumped for fulfilling the plan no matter what the cost."

The blame for such decisions should not be laid solely at the feet of the managers. The criminal US economic blockade, which forced Cuba's integration into the Soviet bloc's economic system (COMECON), gave the country no choice but to apply eastern Europe's resource- and energy-squandering technologies.

Cuba's insertion into the COMECON system retarded the growth of environmental consciousness. Miguel Limia David, a senior researcher with Cuba's Ministry of Science, Technology and Environment (CITMA), has stressed "the predominance of an instrumentalist and personally irresponsible attitude to the use, enjoyment and disposal both of natural as well as socially created resources". Why? For years "we basically aimed at producing more wealth and raising consciousness without paying appropriate attention to the costs of producing that wealth".

Special period

However, even before the 1989-91 collapse of the Soviet bloc threw Cuba's model of highly mechanised agriculture into crisis, problems such as growing pesticide resistance and soil erosion had led to the development of alternatives. In the 1980s, some US$12 billion was devoted to training specialists and developing infrastructure in the areas of biotechnology, health sciences, computer hardware and robotics.

This timely move ensured that when imports of fertiliser, machinery and spare parts fell by 80%, the country was able to devote its scientific knowledge and agricultural research infrastructure to the largest-ever conversion from conventional agriculture to organic or semi-organic farming. This proved vital to maintaining food supplies in very hard times.

However, the 1990s came with many severe environmental problems intact, as identified in the 1997 National Environment Strategy:

- Continuing large-scale soil degradation – erosion, bad drainage, salinity, soil acidity and compacting;
- the deterioration of health and environment conditions in cities and towns, due to a fall in spending on housing and urban infrastructure;
- fresh and salt water pollution that was undermining fishing,

agriculture and tourism, as well as natural ecosystems;

- selective deforestation, which damaged soils, water tables and fragile ecosystems; and
- loss of biological diversity.

The concessions that Cuba has had to make to survive in the capitalist world – such as a large increase in joint ventures in industries like tourism – brings new stresses. Similarly, the growth in numbers of self-employed people and small farmers also threatens to boost environmental decline.

Can Cubans solve their environmental problems? Cuba has the great advantage of having faced facts: the fundamental enemy of global sustainability is capitalism's production for private profit. Capitalism cannot survive without constantly regenerating an anti-environmental and consumerist ethic, no matter what greenwashing corporations say.

As Delgado Díaz explains: "As a spiritual phenomenon, capitalism has produced ways of viewing life and has equipped modern man and woman with an ethical outlook that is incompatible with the solution of the environmental problem that science has advanced as technically viable."

Energy specialist Hector Eugenio Pérez de Alejo Victoria notes that it is vital not to leave the definition of key ecological terms like "eco-efficiency" to promoters of the capitalist market. "The search for a definition is subject to great threats, one of which is the continual propaganda of the international media as to the benefits of consumerism, where a satisfied client is supposedly to be found at the end of every chain. In reality, consumerism is nothing more than an infinite cycle of dissatisfactions; satisfaction for a short period of time and almost immediately more dissatisfaction. It is a sort of drug addiction and produces the greater part of the global environmental disaster."

Humanity-nature relationship

Cuban ecological thinking stresses that the global environmental crisis and the world's social and economic crises are interrelated, in particular through the way the "North" exploits the countries of the "South". As Garrido Vázquez notes: "It is impossible to conceive of

110

sustainable development without resolving beforehand the problems of extreme poverty, which are nothing but the results of centuries of colonial domination and exploitation, and which have re-emerged in recent times through the application of neoliberal policies."

A point of reference is the writings on the humanity-nature relationship by Cuba's national hero and martyr, José Martí. These, in the words of Limia David, "refer to the need to develop a harmonious relationship with the universal conditions of life, with `first nature', as well as to build an ordered, pure and cultured `second nature'".

A succinct expression of this outlook came in Fidel Castro's speech to the 1992 Rio Earth Summit, and has since been matched by a rapid increase in environmental laws and projects within Cuba. Between 1992 and 1998, the National Assembly of People's Power amended the Cuban constitution to entrench the concept of sustainable development; the National Environment and Development Program was developed (outlining the path Cuba would take to fulfil its obligations under the Rio summit's Agenda 21); CITMA was established; an overarching environment law passed; and a national environment strategy was launched.

Other major initiatives included a national strategy for environmental education; a national program of environment and development; projects for food production via sustainable methods and biotechnological and sustainable animal food, as well as a national scientific technical program for mountain zones and a national energy sources development program. Each of these programs is composed of smaller projects and initiatives involving local communities, People's Power bodies, universities, schools and mass organisations.

What has been achieved? There have been gains in health, access to water and electricity, education and land reform, which according to orthodox classification methods are not "environmental" but without which no real advances against environmental degradation are thinkable.

Such gains would never be realised if Cuba reverted to capitalism and was obliged, for example, to pay the US$100 billion debt that Washington estimates Cuba owes for private property expropriated by the revolution. As one environmentalist put it: "The foremost environmental problem we have is making sure we don't fall into the hands of the empire."

Cuba's highly educated people, of whom more than half a million are university graduates, are an invaluable resource base for recent advances such as the conversion to organic agriculture, the thorough surveying of its ecosystems and energy and resource base, the completion of a national biodiversity study, improved methods of water and soil management, and the application of new technologies for treating waste.

Renewable energy and alternative housing

Two fields in which Cuba is making headway against the odds are renewable energy and alternative housing.

Two issues that have focused increased attention on alternative energy are Cuba's high level of dependency on oil imports (around 10 million tonnes annually before 1989) and the fact that its first nuclear reactor has still to come on line, even though work began in the late 1970s.[78]

According to Pérez de Alejo Victoria, the Development Program of National Energy Sources is putting maximum effort into developing energy systems based on sugar cane residues (bagasse), wind farms, micro hydroelectricity plants, solar and photovoltaic technologies as well as on Cuba's unexploited oil reserves.

Cuba's energy goals have been made more difficult by the elimination of some potential energy sources: peat reserves are to be left untouched until environmentally benign methods of peat-burning can be developed and in 1998 the National Assembly of People's Power suspended the construction of Toa-Doaba hydroelectric project, which would have flooded an ecosystem as rare and beautiful as that of Tasmania's Franklin River.

So far, the energy program can boast the generalised usage of bicycles, the development of kerosene substitutes for cooking, the conversion of boilers to enable straw to be burnt as fuel and the increased use of biogas.

The most promising potential energy source is bagasse. With existing technology, Cuba's annual production of 4.3 million tonnes of sugar cane biomass could reduce oil dependency by 700,000 tonnes. If Cuba can gain access to new Brazilian technology which can gasify sugar cane biomass, the country could increase electricity output per biomass unit by up to 10 times — a huge step forward in reducing oil dependency.

During the 1970s and 1980s, Cuba met its relentlessly rising housing demand by building Soviet-style concrete blocks of flats as rapidly as possible. The enforced end of this model of housing development brought some benefits which in the medium term promise more human-scale, environmentally benign housing.

The non-government organisation Habitat-Cuba is devoted to producing a sustainable housing model that recognises that the concrete required for Cuba's standard housing stock has come at a high (and unaccounted for) cost in terms of greenhouse gas emissions and that the passive acceptance of the standard model has led to bureaucratic blindness and indifference towards alternative building materials in which Cuba is rich.

At the same time, the collapse of housing investment during the special period had seen a rise in the number of unhealthy suburbs, especially in older urban areas. This is an urgent challenge to build environmentally sustainable, healthy settlements, basing design, techniques and execution on consultation with local communities, sympathetic architects and other professionals as well as with the relevant ministries.

Habitat-Cuba has developed bamboo as a housing construction material, as well as the introduction of mud-brick techniques – in the face of initial scepticism by a local community who thought they were being returned to stone-age life! Like CubaSolar, an NGO specialising in alternative energy, Habitat-Cuba has built scores of successful projects across the island as well as having provided training in alternative construction techniques.[79]

Towards a lasting solution

Despite such advances Cuba's environmentalists do not underestimate the difficulties their country's environment faces. Delgado Díaz

113

points out that "it is extraordinarily difficult to break the vicious circle of underdevelopment, environmental degradation and poverty. Phenomena of this type impose an individual economic dynamic that is often resolved at the expense of the environment."

What are the prospects? Pérez de Alejo Victoria said that "the environmental realities are pretty unflattering, especially as regards renewable energy, which obliges me to be tactically pessimistic, even if from the strategic point of view I view the future with optimism".

Limia David is less hopeful. He thinks environment policy can only work to its full potential if Cuban society overcomes the indifference generated by its paternalistic heritage, conquering "the unsatisfactory degree of involvement of the direct producers in the means of production, that is, the inadequate linkage between everyone's way of life and the final results of the production process".

For David, Cuba's acute environmental problems cannot be solved by political will alone, necessary and important though that is: "They essentially demand not a new attitude on the part of policy generated by the state and the entire political system, but one that arises from the ordinary people, from the local communities and specific labour collectives. It is critical to develop a feeling of responsible ownership when faced with the universal bases of life."

However, Modesto Fernández Díaz-Silveira, a CITMA specialist in the management of environment policy, is more confident: "The sustained economic recovery and institutional changes that are taking place in Cuba provide a solid basis that allow us to advance with optimism in the application of our environmental policy, the norms and methods of application of which will take us to a higher stage in the protection of the environment and the rational use of natural resources."

The main factor behind this confidence is the mass participation and revolutionary commitment of Cuba's people and communities in implementing environment policy, an ingredient that no capitalist society can match. Even while Cuba still lags in making use of many of the tools available to capitalist governments (eco-taxes, environmentally adjusted national accounting), participatory democracy gives Cuba the chance to advance towards sustainability while in the rest of the Third World the environment collapses.

This is especially so when combined with the Cuban political system's capacity to implement integrated plans involving all "players" and its desire to educate its people in humanist and environmental values.

There is a broad debate on the island about how to involve the mass of people in the battle for environmental sustainability. That is far more inspiring and hopeful than an environment policy which consists of Dodgy Brothers flogging us shares in tax-deductable eucalypt plantations.

Part four: growth and nature

An ecosocialist manifesto

Introduction

The idea for this ecosocialist manifesto was jointly launched by Joel Kovel and Michael Löwy, at a September, 2001, workshop on ecology and socialism held at Vincennes, near Paris. We all suffer from a chronic case of Gramsci's paradox, of living in a time whose old order is dying (and taking civilization with it) while the new one does not seem able to be born. But at least it can be announced. The deepest shadow that hangs over us is neither terror, environmental collapse, nor global recession. It is the internalized fatalism that holds there is no possible alternative to capital's world order. And so we wished to set an example of a kind of speech that deliberately negates the current mood of anxious compromise and passive acquiescence. This manifesto nevertheless lacks the audacity of that of 1848, for ecosocialism is not yet a spectre, nor is it grounded in any concrete party or movement. It is only a line of reasoning, based on a reading of the present crisis and the necessary conditions for overcoming it. We make no claims of omniscience. Far from it, our goal is to invite dialogue, debate, emendation, above all, a sense of how this notion can be further realized. Innumerable points of resistance arise spontaneously across the chaotic ecumene of global capital. Many are immanently ecosocialist in content. How can these be gathered? Can we envision an "ecosocialist international?" Can the spectre be brought into being?

Manifesto

The twenty-first century opens on a catastrophic note, with an unprecedented degree of ecological breakdown and a chaotic world order beset with terror and clusters of low-grade, disintegrative warfare that spread like gangrene across great swathes of the planet – viz, central Africa, the Middle East, Northwestern South America – and reverberate throughout the nations. In our view, the crises of ecology and those of societal breakdown are profoundly interrelated and should be seen as different manifestations of the same structural forces.

The former broadly stems from rampant industrialization that overwhelms the earth's capacity to buffer and contain ecological destabilization. The latter stems from the form of imperialism known as globalization, with its disintegrative effects on societies that stand in its path. Moreover, these underlying forces are essentially different aspects of the same drive, which must be identified as the central dynamic that moves the whole: the expansion of the world capitalist system.

We reject all euphemisms or propagandistic softening of the brutality of this regime: all greenwashing of its ecological costs, all mystification of the human costs under the names of democracy and human rights. We insist instead upon looking at capital from the standpoint of what it has really done. Acting on nature and its ecological balance, the regime, with its imperative to constantly expand profitability, exposes ecosystems to destabilizing pollutants, fragments habitats that have evolved over aeons to allow the flourishing of organisms, squanders resources, and reduces the sensuous vitality of nature to the cold exchangeability required for the accumulation of capital.

From the side of humanity, with its requirements for self-determination, community, and a meaningful existence, capital reduces the majority of the world's people to a mere reservoir of labor power while discarding much of the remainder as useless nuisances. It has invaded and undermined the integrity of communities through its global mass culture of consumerism and depoliticization. It has expanded disparities in wealth and power to levels unprecedented in human history. It has worked hand in glove with a network of corrupt and subservient client states whose local elites carry out the work of repression while sparing the center of its opprobrium. And it has set going a network of transtatal organizations under the overall supervision of the Western powers and the superpower United States, to undermine the autonomy of the periphery and bind it into indebtedness while maintaining a huge military apparatus to enforce compliance to the capitalist center. We believe that the present capitalist system cannot regulate, much less overcome, the crises it has set going. It cannot solve the ecological crisis because to do so requires setting limits upon accumulation—an unacceptable option for a system predicated upon the rule: Grow or Die! And it cannot solve the crisis posed by terror and other forms of violent rebellion because to do so would mean abandoning the logic of empire, which would impose unacceptable limits on growth and the whole "way of life" sustained by empire.

Its only remaining option is to resort to brutal force, thereby increasing alienation and sowing the seed of further terrorism . . . and further counter-terrorism, evolving into a new and malignant variation of fascism. In sum, the capitalist world system is historically bankrupt. It has become an empire unable to adapt, whose very gigantism exposes its underlying weakness. It is, in the language of ecology, profoundly unsustainable, and must be changed fundamentally, nay, replaced, if there is to be a future worth living. Thus the stark choice once posed by Rosa Luxemburg returns: Socialism or Barbarism!, where the face of the latter now reflects the imprint of the intervening century and assumes the countenance of ecocatastrophe, terror counterterror, and their fascist degeneration.

But why socialism, why revive this word seemingly consigned to the rubbish-heap of history by the failings of its twentieth century interpretations? For this reason only: that however beaten down and unrealized, the notion of socialism still stands for the supersession of capital. If capital is to be overcome, a task now given the urgency of the survival of civilization itself, the outcome will perforce be "socialist", for that is the term which signifies the breakthrough into a post-capitalist society. If we say that capital is radically unsustainable and breaks down into the barbarism outlined above, then we are also saying that we need to build a "socialism" capable of overcoming the crises capital has set going. And if socialisms past have failed to do so, then it is our obligation, if we choose against submitting to a barbarous end, to struggle for one that succeeds. And just as barbarism has changed in a manner reflective of the century since Luxemburg enunciated her fateful alternative, so too, must the name, and the reality, of a socialism become adequate for this time.

It is for these reasons that we choose to name our interpretation of socialism as an ecosocialism, and dedicate ourselves to its realization.

Why Ecosocialism?

We see ecosocialism not as the denial but as the realization of the "first-epoch" socialisms of the twentieth century, in the context of the ecological crisis. Like them, it builds on the insight that capital is objectified past labor, and grounds itself in the free development of all producers, or to use another way of saying this, an undoing of the separation of the producers from the means of production. We understand that this goal was not able to be implemented by first-epoch socialism, for reasons too complex to take up here, except to summarize as various effects of underdevelopment in the context of hostility by existing capitalist powers. This conjuncture had numerous deleterious effects on existing socialisms, chiefly, the denial of internal democracy along with an emulation of capitalist productivism, and led eventually to the collapse of these societies and the ruin of their natural environments. Ecosocialism retains the emancipatory goals of first-epoch socialism, and rejects both the attenuated, reformist aims of social democracy and the productivist structures of the bureaucratic variations of socialism. It insists, rather, upon redefining both the path and the goal of socialist production in an ecological framework. It does so specifically in respect to the "limits on growth" essential for the sustainability of society. These are embraced, not however, in the sense of imposing scarcity, hardship and repression. The goal, rather, is a transformation of needs, and a profound shift toward the qualitative dimension and away from the quantitative. From the standpoint of commodity production, this translates into a valorization of use-values over exchange-values—a project of far-reaching significance grounded in immediate economic activity.

The generalization of ecological production under socialist conditions can provide the ground for the overcoming of the present crises. A society of freely associated producers does not stop at its own democratization. It must, rather, insist on the freeing of all beings as its ground and goal. It overcomes thereby the imperialist impulse both subjectively and objectively. In realizing such a goal, it

119

struggles to overcome all forms of domination, including, especially, those of gender and race. And it surpasses the conditions leading to fundamentalist distortions and their terrorist manifestations. In sum, a world society is posited in a degree of ecological harmony with nature unthinkable under present conditions. A practical outcome of these tendencies would be expressed, for example, in a withering away of the dependency upon fossil fuels integral to industrial capitalism. And this in turn can provide the material point of release of the lands subjugated by oil imperialism, while enabling the containment of global warming, along with other afflictions of the ecological crisis.

No one can read these prescriptions without thinking, first, of how many practical and theoretical questions they raise, and second and more dishearteningly, of how remote they are from the present configuration of the world, both as this is anchored in institutions and as it is registered in consciousness. We need not elaborate these points, which should be instantly recognizable to all. But we would insist that they be taken in their proper perspective. Our project is neither to lay out every step of this way nor to yield to the adversary because of the preponderance of power he holds. It is, rather, to develop the logic of a sufficient and necessary transformation of the current order, and to begin developing the intermediate steps towards this goal. We do so in order to think more deeply into these possibilities, and at the same moment, begin the work of drawing together with all those of like mind. If there is any merit in these arguments, then it must be the case that similar thoughts, and practices to realize these thoughts, will be co-ordinately germinating at innumerable points around the world. Ecosocialism will be international, and universal, or it will be nothing. The crises of our time can and must be seen as revolutionary opportunities, which it is our obligation to affirm and bring into existence.

Joel Kovel and Michael Löwy   *Paris, Sept 2001*

If you support the argument of the Manifesto,
and would like to be kept up-to-date on it,
and/or contribute to its development,
send an e-mail to jkovel@prodigy.net

Notes on contributors

John Bellamy Foster is an editor of *Monthly Review*. He is a professor in sociology at the University of Oregon. He is the author or numerous works on Marxism including *Marx's Ecology* (200: Monthly Review Press, New York).

Alice Cutler is an activist in the movement against climate change.

Jane Kelly is a supporter of *Socialist Resistance*. She edits *Socialist Outlook*, the quarterly magazine of the International Socialist Group, and is a contributor to *International Marxist Review*.

Joel Kovel is an American politician, academic, writer and ecosocialist; he is a member of the Green Party of the US (GPUS)

Michael Löwy is Research director of Sociology at the Centre National de la Recherche Scientifique (CNRS) in Paris, and a member of the LCR (Ligue Communiste Révolutionnaire – French section of the Fourth International)

François Moreau, 1956-1993. A trained economist and professor at the University of Toronto, he was the author of three books on the Quebec economy and of many contributions to specialized journals and collective works. He was a member of the Fourth International in the Canadian State, and a member of the International Executive Committee of the Fourth International.

Sheila Malone is a supporter of *Socialist Resistance* and an activist in the Campaign against Climate Change. She has been a member of the editorial board of *Labour Focus on Eastern Europe* and is on the editorial board of *Socialist Outlook*.

Dick Nichols is the managing editor of *Seeing Red*, the magazine initiated by the Australian Socialist Alliance. He has also edited the analysis of the Cuban revolution of the Democratic Socialist Perspective, an affiliate organization of the Alliance.

Daniel Tanuro is an environmentalist and the ecological correspondent of *La Gauche*, the French-language newspaper of the Socialist Workers Party (POS/ SAP, Belgian section of the Fourth International).

Phil Ward is a supporter of *Socialist Resistance* and a long-standing militant of the Fourth International. He serves on the climate change commission of the International Socialist Group.

Further Reading

Joel Kovel, *The Enemy of Nature: The End of Capitalism or the End of the World,* Zed Books, 2002

John Bellamy Foster:
Marx's Ecology: Materialism and Nature, Monthly Review Press, 2000

Ecology against Capitalism, Monthly Review Press, 2002

'A Planetary Defeat: the Failure of Global Environmental Reform', *Monthly Review,* v. 54, no. 8, January 2003
http://www.monthlyreview.org/0103jbf.htm.

'Ecology Against Capitalism', *Monthly Review,* v. 53, no. 5, October 2001.
http://www.monthlyreview.org/1001jbf.htm.

'The Scale of Our Ecological Crisis', *Monthly Review,* v. 49, no. 11, April 1998.   http://www.monthlyreview.org/498jbf.htm.

'Capitalism's Environmental Crisis: Is Technology the Answer?', *Monthly Review,* v. 52, no. 7, December 2000.
http://www.monthlyreview.org/1200jbf.htm.

'Marx's Ecology in Historical Perspective', *International Socialism Journal,* no.96, 2002
http://pubs.socialistreviewindex.org.uk/isj96/foster.htm.

'Ecosocialisme ou Barbarie', *Critique Communiste,* 177, Octobre 2005.
Available from Librairie La Breche, 27, rue Taine, 75012 Paris
Email: contact@la-breche.com

Other books from Socialist Resistance

Ron Ridenour, *Cuba beyond the crossroads,* Socialist Resistance, 2006.

Celia Hart (Edited by Walter Lippman), *It's Never Too Late To Love Or Rebel,* Socialist Resistance, 2006.

To order send a cheque or international money order for £10 per book, payable to Socialist Resistance to PO Box 1109, London, N4 2UU

Footnotes

1   Concentrations of CO_2, CH4 (methane) and N_2O (nitrous oxide), three of the main greenhouse gases, have increased respectively by 30%, 145% and 15% from 1750.

2   The IPCC was set up in 1988 by the UN Environment Programme and the World Meteorological Organization. Its scientific analyses have authority. See the history by Nicolas Chevassus-au-Louis, La Recherche, number 370, December 2003.

3   The Guardian, July 28, 2003. The parallel with terrorism has been taken up by David King, scientific adviser to Tony Blair.

4   "Developing" countries have no objectives within the framework of the first period 2008-2012.

5   Around 50% of US electricity power stations are fuelled by coal, and four fifths of the generation capacity that the country will need in 2010 was already installed in 2000.

6   The US and Russia occupy first and fourth place respectively in the league table of countries responsible for emissions of greenhouse gases. The US, with 5% of the world population, uses 25 % of the world's energy resources.

7   New York Times, December 3, 2003. (Note added in 2006: Russia's parliament voted to join the Kyoto protocol in October 2004, in exchange for EU support for Russia joining the World Trading Organisation. This meant that the Kyoto Protocol became legally binding and the 34 industrialised countries in it will be penalised if they don't meet their targets. Russia's target is to keep its emissions at 1990 levels. Although rising, emissions are still at least 20% below 1990 levels, due to the economic collapse after the end of the Soviet Union in 1991.)

8   See David Victor, The collapse of the Kyoto protocol and the struggle to slow global warming, Princeton University Press, 2001.

9   Inside the EU the policies currently followed by the member states will lead to a reduction of emissions in Europe by only 0.5% in 2010 in relation to 1990. Taking into account measures planned but not yet applied, the reduction will be at most 7.2 % and more probably 5.1% (EEA, "Greenhouse gas emission trends and projections in Europe" 2003. Environment Issue Report 36). (Note added in 2006: EU reductions so far are 1.7% and the 8% target is unlikely to be met.)

10 New York Times, December 4, 2003.

11 Idem.

12 IPCC, Third evaluation report, report of Working Group I, technical summary.

13 The Sahara is undoubtedly the result of a crossing of the threshold of this kind. We know now that its formation 5,000 years ago took only a few centuries.

14 Even without massive melting ocean levels are rising because of expansion of water masses. According to the IPCC's third assessment report, several tens of millions of people will be displaced by 2100. With a rise of one metre, nearly 25% of the population of Vietnam would have to be evacuated. The melting of the Arctic ice has begun. Melting in the Antarctic does not seem significant - happily, because the disappearance of the Southern icecap would raise sea levels by around 63 metres!

15 Larry Lohman, "Democracy or Carbocracy? Intellectual corruption and the future of the climate debate". Corner House Briefing number 24.

16 In addition to fossil fuels, permanently frozen areas of land are big reservoirs of carbon that are at the moment removed from the cycle. This carbon could be freed if there was a melting, which is an example of the kind of "feedback effect" possible in the climatic system.

17 Third evaluation report, report of Working Group I, technical summary, pp. 60-61.

18 Third evaluation report, report of Working Group III, technical summary, p. 40. This rate would only be reached if all the surfaces deforested for two centuries were reforested by 2100, which is very unlikely.

19 Kyoto Protocol, article 3.

20 Some studies indicate that the "sinks" could be transformed into "sources". In the tropical forests, for example, the increase in the concentration of CO_2 would favour the proliferation of creepers in such quantity that the weakened ecosystem would emit CO_2 instead of capturing it.

21 Financial Times, June 25, 2003.

22 "Greenhouse gas emission trends and projections in Europe 2003". EEA, Environment Issue Report 36.

23 VOA News, October 22, 2003.

24 David Victor, The collapse of the Kyoto Protocol and the Struggle to Slow Global Warming, Princeton University Press, 2001.

25 Rather than reducing emissions, some technocrats are trying to perfect systems for deep burial of CO_2. Read "Putting Carbon in its Place", Business Week, October 29, 2003.

26 This perverse effect of the CDM on the possibilities of development of the countries of the South is called the "low hanging fruits" effect.

27   The proportion in the USA is only 2%.

28   Eurostat and COM (2002) 162 final. "Décision du Parlement et du Conseil: Une énergie intelligente pour l'Europe".

29   See Commission document: http://ec.europa.eu/environment/climat/pdf/ climate_focus_en.pdf "EU Focus on Climate Change" 2002

30   Analysis of the EU ecoindustries, their employment and export potential: Commission document.

31   "Global Warming: Bush's Double Blunder". Business Week, April 9, 2001.

32   "Global Warming: Has Bush on the Hot Seat". Business Week, op. cit.

33   An in-depth analysis of Plantar can be found on the website of FERN at www. fern.org.

34   Heidi Bachram, 'Climate Fraud and Carbon Colonisation: The New Trade in Greenhouse Gases', Capitalism, Nature, Socialism, Vol 15, No.4 (December 2004)

35   Much of the information in the section above is from the American Institute of Physics web site: http://www.aip.org/history/climate/index.html. Also Scientific American, Vol 260, No4, pp18-26, 1989; and Environmental Science and Technology, Vol 3, No 11, pp1162-1169, 1969.

36   Because of this 'thermal lag', some models estimate that current GHG levels should lead to an average rise in global temperatures of 4oC since 1860, not the 0.6cC that has occurred so far.

37   New Scientist, 3 December 2005, pp36-41. Other information on developments in 2005 is from various articles in Nature, the Environmental News Service and the Science and Development Network (SciDevNet).

38   The 1997 Kyoto Protocol, called for a mere cut of 5.2% in GHG emission levels for industrialised (Annexe B) countries from between 1990 and 2010. By 2002, global emissions had risen 11% including a rise in Annexe B countries. In the UK, the decline in emissions preceded 1997 and was due to the 'dash to gas', replacing coal for electricity generation. Little effort was put into the areas that are traditionally difficult – household energy use and transport. There is now a 'dash to coal', as gas prices rise and coal power stations, mothballed in the 1990s, are brought back into use.

39   The Montreal 'agreement' of December 2005, which brought Margaret Beckett to tears of ecstasy and was hailed by Greenpeace as 'a big step forward', opened a 'dialogue' and explicitly excluded 'negotiations leading to new commitments' after the Kyoto period ends. Bush's resistance is usually attributed to his being in hock to the oil companies: one commentator has suggested that it is

because some studies suggest US power would be greater after a global climate catastrophe. This is certainly the view of the notorious Pentagon document 'An Abrupt Climate Change Scenario and Its Implications for United States National Security'.

40 *'We must cut demand to have any hope of solving the energy crisis', Guardian, November 29th 2005. http://www.guardian.co.uk/Columnists/Column/0,,1653215,00.html*

41 *Nature, 13th June 2003.*

42 *See 'The nuclear option: A solution to global warming?', in Socialist Outlook 6, Spring, 2005, for a more detailed critique of nuclear power.*

43 *For a graphic account of the repressive activities of the police after Katrina see http://www.counterpunch.com/bradshaw09062005.html Mike Davis shows how a government can ignore potential disasters staring them in the face, even when the (financial) cost of doing so is huge: http://mondediplo.com/2005/10/ 02katrina. Davis has also documented the attempts to 'cleanse' New Orleans of a large number of its black and working class residents since Katrina: 'Gentrifying Disaster', Mother Jones, 25th October 2005 and 'Hurricane Gumbo', The Nation, 7th November 2005 (both on the web).*

44 *Thanks are given for technical information to Dr Ian Fairlie, a member of SERA (Socialist Environment and Resources Association).*

45 *See www.dti.gov.uk/expenditureplan/report2004*

46 *Intergovernmental Panel on Climate Change*

47 *http://www.stopclimatechaos.org*

48 *http://www.scoop.co.nz/stories/WO0602/S00497.htm*

49 *http://www.dti.gov.uk/energy/review/energy_review_intro.pdf*

50 *New Scientist, 18th January 2003*

51 *International Atomic Energy Authority*

52 *BNFL and Westinghouse are the largest foreign investors in the project, which is run by the S African power company ESKOM.*

53 *Safe New Generation Nuclear Power? http://www.i-sis.org.uk/SNGNP.php*

54 *In http://www.stormsmith.nl/, it is argued that nuclear power stations could end up producing more GHGs than efficient gas-powered stations, while http://www.nuclearinfo.net/Nuclearpower claims that, over a 40-year period, a particular nuclear plant produces 93 times the energy as it consumes.*

55 *At great cost to 'the taxpayer' it should be added, along with considerable safety risks. In May 2005, BNFL announced that it had had at its THORP reprocessing plant in Sellafield an acid leak over the previous 4 to 8 months. In*

*this time, sufficient plutonium-laden waste to construct 20 nuclear warheads,
and 20 tonnes of uranium, fell into a concrete-lined cell.
See http://www.i-sis.org.uk/ESIGW.php. BNFL is a state company (we
described it incorrectly in Socialist Outlook 8), currently being sold off.
Its plant company, Westinghouse, went to Toshiba and BNG America
(reprocessing and clean-up) to Energy Solutions. BNG (UK) is up for grabs.*

56 *New Scientist, 25/2/06, reports research into "hydrogen farms" consisting
of algae, genetically engineered to photosynthesise the gas. Another scheme
is to use algae to photosynthesise captured carbon dioxide from power
stations to make biodiesel and algal "cake" (ISIS - http://www.i-sis.org.
uk/GAFCCAB.php). Both look interesting, but (as ever) as an alternative
to energy conservation and social change, questions of scale and consequent
environmental damage come into play. ISIS says algal diesel requires 40,000
square km of desert to satisfy (current) US demand. That is one sixth of the
UK's total land area, and probably doesn't include spaces between plants,
service roads etc, which would at least double the area. Hydrogen damages the
ozone layer and burning of all fuels creates nitrogen oxides. Human activity
has now completely overwhelmed the natural nitrogen cycle and is causing
major pollution problems (New Scientist, 21/1/06).*

57 *Paul Roberts: The End of Oil, Bloomsbury, 2004. Roberts shows how
government deregulation of the energy supply companies has increased the
incentives to sell energy (renewable or not) rather than conserve.*

58 *Some, mostly bourgeois, commentators argue that these changes are difficult
to predict or, as in the case of Piers Corbyn, a radical physicist, that they
are the result of changes in the sun. It seems sensible, however, to adopt a
precautionary principle rather than wait to see what happens.*

59 *While protests against the use of asbestos have reduced its use in the rich
industrialised countries, many are now dying from asbestosis and related lung
diseases, because the cancers it produces are slow in developing. Its use is still
rising in neo-colonial countries.*

60 *Engels, (1845) 1977, The Condition of the Working Class in England,
Lawrence & Wishart, London p. 45.*

61 *Engels, (1845), p. 81*

62 *Engels, (1845), p 85*

63 *Colin Hines and Caroline Lucas, 'Time to replace Globalisation: A Green
Localist Manifesto for the World Trade Organisation Ministerial', p 21.
Replace_globalisation. pdf*

64 *Ibid p. 22*

65 *http://www.garretthar dinsociety.org/articles/articles.html*
    Also interesting is Hardin's later essay 'Lifeboat Ethics: the case against
    helping the poor', available at the same web site. Another early theorist of
    environmental problems, Kenneth Boulding, argued in 1966 in 'The economics
    of the coming spaceship earth', (a term which he coined) that price mechanisms
    were insufficient to achieve the change from the 'cowboy economy' to the
    'spaceman economy' (http://dieoff.org/page160.htm).

66 *This may partly explain the Green's extremely sectarian response to the*
    RESPECT project originally initiated by Monbiot/Yacoub and Galloway.

67 *The number of cars is set to double by 2020.*

68 *Hans Magnus Enzensberger, ' A Critique of Political Ecology', New Left*
    Review, I/84, March-April 1974, pp. 3–31.
    This essay is also in Raids and Reconstructions, Pluto, 1976 pp 253-295.

69 *This is a revised version of a keynote address delivered to the Critical*
    Management Studies section of the Academy of Management in Honolulu,
    Hawaii on August 8, 2005.

70 *The authors of the Global Scenario Group's Great Transition report are Paul*
    Raskin, Tariq Banuri, Gilberto Gallopín, Pablo Gutman, Al Hammond, Robert
    Kates, and Rob Swart.

71 *Much of Marx's analysis in Capital is concerned with where m or surplus value*
    comes from. To answer this question, he argues, it is necessary to go beneath the
    process of exchange and to explore the hidden recesses of capitalist production —
    where it is revealed that the source of surplus value is to be found in the process
    of class exploitation.

72 *http://www.monthlyreview.org/0504editors.htm*

73 *http://www.monthlyreview.org/1204jbfclark.htm*

74 *On Marx's relation to Epicurus see John Bellamy Foster, Marx's Ecology (New*
    York: Monthly Review Press, 2000).

75 *This is an edited version of a resolution on climate change adopted by the British*
    section of the Fourth International in April 2006.

76 *The Fourth Internatiomal's international committee adopted this resolution at its*
    meeting in February 2006.

77 *This is an extract from the resolution of the 15th world congress of the Fourth*
    International on 'Socialism and Ecology'

78 *Cuba had started the construction of two reactors at Juragua, in Cienfuegos.*
    Work was suspended in 1992 following the collapse of the Soviet Union. Cuba

129

has now abandoned its nuclear power program.

79   *Habitat-Cuba no longer exists. Some observers feel that it lost out in a power
      struggle with the Ministry of Housing. There's an understandable history of
      tensions between NGOs and the ministries, with many in power regarding the
      NGOs as potential trojan horses for imperialism. This is not pure paranoia, of
      course, but is often unfortunate.*

Printed in the United Kingdom
by Lightning Source UK Ltd.
116165UKS00001B/89